START TO END ON HOW TO BUILD A HOUSE

BY VIRAJ PATIL

■ ■ ■ ■ ■ ■ ■ ■ ■ ■

Copyright©2018 by **PATIL VIRAJ PARESH MRUDULA**

All rights reserved. This book or any part of this book is not allowed to be reproduced, distributed or transmitted in any forms including PDFs, audio, video or other digital or physical methods without the written permission of the author

E-Mail – **patilviraj1997@gmail.com**

ISBN - 9781977053329

Publishing date – 30 January 2018

■ ■ ■ ■ ■ ■ ■ ■ ■ ■

DEDICATION

Dedicated to my dad **MR.PARESH RAMESH PATIL** for inspiring me with his constant hard work to run our family without a single knock on our comfort

Dedicated to my mom **MRS. MRUDULA PARESH PATIL** for sacrificing and investing her whole life for my growth and development in life

Dedicated to my Grandma **MRS. BHAVNA RAMESH PATIL** for all the measureless love and for binding our entire family in a plate; clearly the glue of the PATIL FAMILY

Dedicated to my grandpa **MR. RAMESH PATIL** for just being present and stimulate good vibes

Finally, dedicated to each and every member of **PATIL FAMILY** and my unalloyed **FRIENDS** for constantly motivating me to do great things in my life

----Start To End On How To Build A House----

PREFACE

After my admission in Civil Engineering College, I used to always wonder what is basically civil engineering or how is a house made?

College books were providing knowledge in parts, and never covered the whole process including important small operations needed for it.

Finally after failing to find such kind of book I decided why not write a book instead. So with the assistance of my dad who holds an experience of 15 years in construction business; wrote this book.

A contractor (builder) knows something that engineers don't, and an engineer knows something that a contractor doesn't. So basically this book is a perfect mixture of both, comprising of every single step starting from the knowledge needed for selecting a plot to the final completion/occupancy certificate.

I made it with an aim of clearing your concepts about what is building a house is all about.

Disclaimer

This book does not cover the designing part of the house.

The process of design involves finding out the sizes of structural members (beams, columns etc) and the proper combination and grades of steel and concrete required to resist the intended load safely and economically.

Table of content

Chapter no.	Title	Page no.
1	Selecting a Plot	9
2	Survey	13
3	Planning	21
4	Necessary paper work	43
5	Foundation	53
6	Plinth beam	88
7	Column	92
8	Walls	104
9	Staircase	124
10	Slabs and Beams	134
11	Plastering	146
12	Plumbing	155
13	Wiring	169

14	Painting	172
15	Flooring	175
16	Doors and Windows	181
17	Completion/Occupancy Certificate.	187

Chapter 1 – Selecting a Plot

The very 1st pace before you conclude anything about what to do and what to build or how your house may look like, is to find an ideal plot.

The selection of plot is very crucial, as it affects the overlying structure and planning of the same. There are numerous factors which takes hold of the ideality of the plot. I have exchanged views on some of the factors below.

Soil. The colour of the soil says it all. Soil which are red, brick red, brown, yellow in colour are good for construction. Dark brown or black colour soil indicates that the soil has high organic matter content and also has greater moisture content, hence making it unfit for construction as it may lack in load bearing capacity.

To be safe and sound the sample of soil is sent for testing in laboratory.

Post Script: Avoid soil with ant farms and worms.

The profile of the plot should be **square or rectangular**, as these shapes allow maximum utilization of land.

While in case of irregular shapes the land is wasted due to its irregularity.

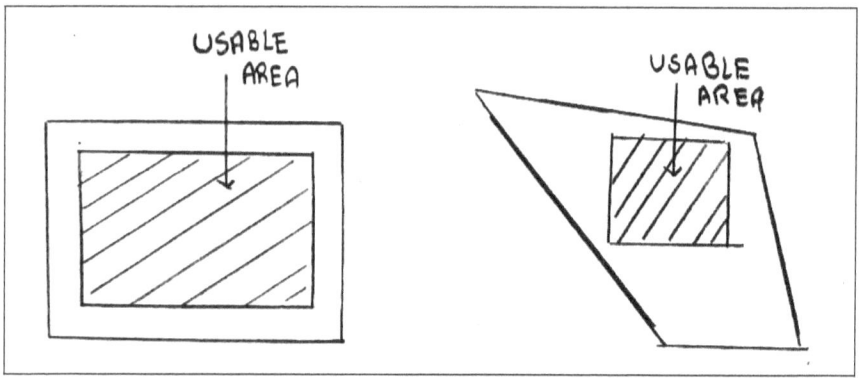

Validate that there is **enough space** available for building after leaving out frontier **for future widening of** adjoining **road**ways.

The **plot** should be **in a developed area** having facilities like, shopping, educational institutions, recreation, hospital and utility service like water supply, drainage system, gas supply, electricity, etc.

Avoid **sandwiched plot.** The plot selected should not be between high rise buildings. This may block the sunlight, breeze as well as view of the occupants.

Accessibility to road. Things get better if we have a constructed road accessible to the plot.

Connectivity to public transport. When we live on a permanent address its profitable to have transport facilities near you. Make sure that the public transport facilities are at a walking distance away from the plot.

Picking the plot in **Residential zone** is a first-class choice, as commercial and industrial zone are not suitable for livelihood.

The **level of the plot** must not be lower than the road level and the level of adjoining plots, as it may cause water logging problems in rainy seasons. The level must be at least equal to the level of the road or higher than that is acceptable.

Huge trees near the plot are unsafe as they are susceptible to thunder lightening. Huge trees don't cause any problem if they are within our plot premise as we can uproot them. But if the trees are in neighbors plot or by any reason we cannot uproot them, then there's a problem.

The plot should be **clear title**. Clear title is the phrase used to state that the owner of real property owns it free and clear of encumbrances. Meaning, free from any disputes and court judgements.

NA land. See to it that the plot you buy is non agricultural land. If not, then you need to transfigure it into NA land.

The above were the prime factors taken into consideration. Not all the factors would be feasible as the final selection of plot falls on the budget of the buyer.

After plot is nominated, it is cleared off by cutting shrubs, uprooting trees, and is leveled up by cutting and filling of necessary portions.

Chapter 2 – Survey

❖ Definition in civil engineering books:

Surveying is the art of determining the relative position of different objects on the surface of the earth by measuring the horizontal distance between them and preparing a map to any suitable scale.

❖ Application of surveying:

1. To draw up a topographical map which shows hills, valleys, rivers, villages, towns, forests, etc. of a country.
2. To prepare a cadastral map. This shows boundaries of each house and property.
3. To prepare engineering map which displays the details of engineering works like roads, railways, reservoirs, canal, etc.

4. To prepare a contour map. This helps in arranging best able to be done routes for transportation.
5. To prepare military map showing roads and railways network with different parts of country. Such maps are prepared for defense of the country. To prepare geological map showing regions and zones of underground resources like mineral, oil, etc.
6. To prepare an archaeological map including places where ancient memorial exist.

But for building a house we should only pivot on point no. 2 that is the cadastral survey. Cadastral surveying is a fragment of surveying which deals with the real property boundaries i.e. the boundaries of the plot.

Cadastral survey is performed so as to check whether your property lines do not move into neighbor's property line or vice versa. Care must be taken while performing because if the surveyor is off by even a small verge and intrude into neighbor's plot, then the neighbor can take you to court. So while you hire a surveyor ensure that they are licensed and have satisfactory insight of the same work.

To originate the survey, you need a reference point to start from. This reference point can be anything like a rock, a tree or a metal pole.

To find the reference point you must have at least one document of the land in which there should be a section named 'legal descriptions' which expound the land and the boundary layout of it.

Another way to locate the reference point is by referring city map. Check with your town or county hall if they have these records with them.

When the reference point is located, the further process starts. The very effective and accurate process to carry out cadastral survey is 'metes and bounds' system. It is explained below.

❖ METES AND BOUNDS SYSTEM

This system works on correlation between the bearing i.e. direction (North, South, East, West) and length i.e. distance to locate the corner points of the plot. The mutuality of the direction and distance is mentioned in the 'legal description' section of the land papers.

Example of a correlation -

'Thence, N45°30'W for a distance of 5meter to a point of line.'

This means we should move in the north- west (NW) direction at an angle of 45°30' and set out 5 meter with the help of measuring tape.

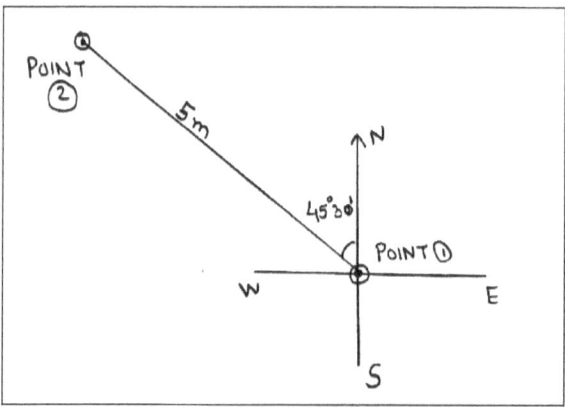

The direction is measured using compass and length is measured using normal tape or chain.

- **Point to remember:**

Angles are measured only with reference to north (N) and south (S) axis.

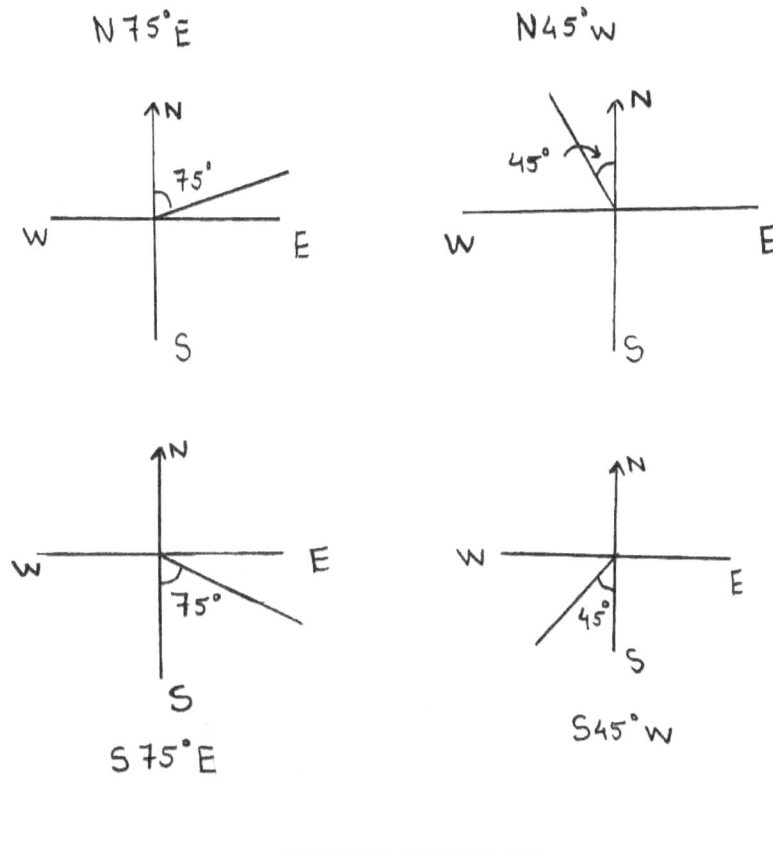

By performing each description we will get the actual boundary line of the plot.

For better understanding an example of legal description and drawing of the plot is given:

From point of beginning, Thence N10°E for a distance of 20meters to a point on a line.

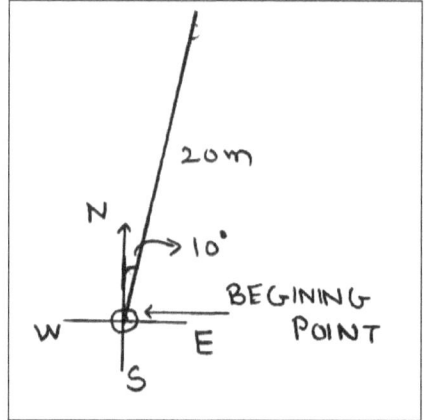

Thence N80°W for a distance of 30m to a point on a line.

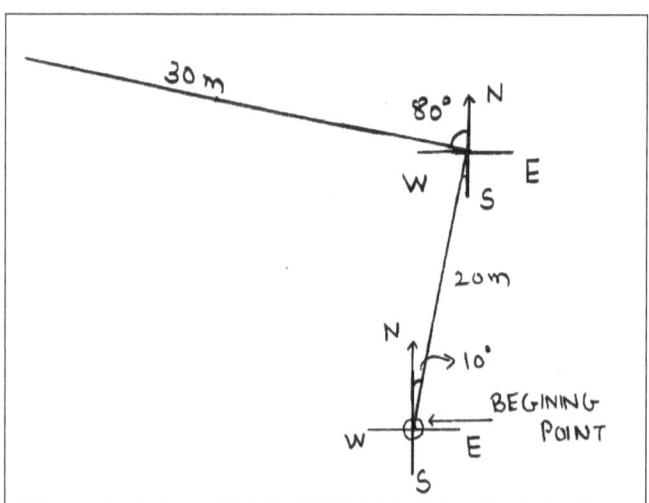

Thence S10W for a distance of 25m to a point on a line.

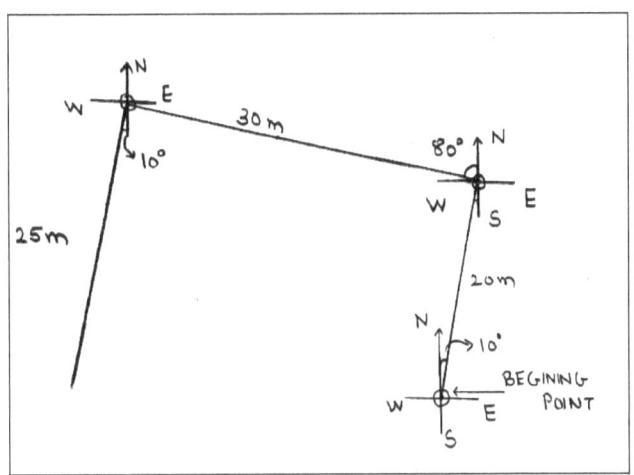

Thence towards the beginning of the point and complete the quadrilateral.

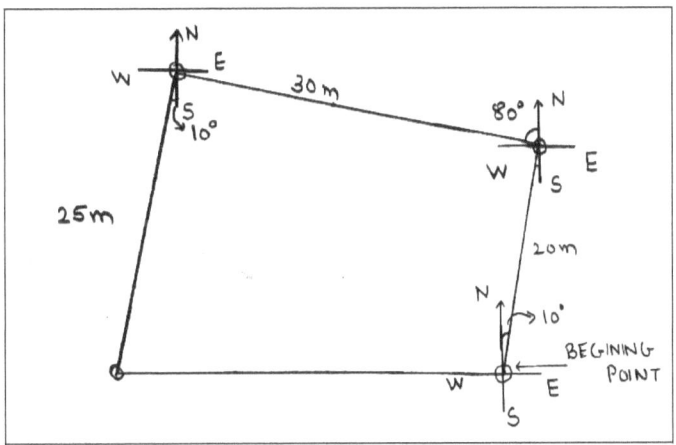

Once the whole survey is done and the clear-cut boundary of plot is investigated, it is vital to circumscribe the plot by fencing it.

Chapter 3 – Planning

After the survey is done and the actual are of the plot is known, we move forward by planning the house you desire to build.

The house plan should be prepared according to the ask, economic status and taste of the owner.

Houses are generally planned and designed based on some rules and regulations. In India it is called as 'Vastu – Shastra', which literally translates to "Science of architecture". Vastu – Shastra is fully based on science and has a scientific reasoning for its every rule. There are myths like "If the entrance does not face the north direction, then the people living in will be unhappy or they will suffer from deadly diseases." etc. etc.

The aspect and reasoning for the same is given below

1] Aspect:

It is the positioning of the rooms in the needed directions. Also involves positioning of doors and windows to enjoy natural gifts like sunshine and breeze to the maximum level.

Sr. no.	Units	Aspect	Remark	Reason
1.	Kitchen	E/NE/ SE	Morning sunlight should invade in.	Morning sunlight kills the germs and bacteria present on kitchen platform. As this unit is used for cooking so hygienic atmosphere should be preserved here.
2.	Bedroom	W/NW /SW	Evening sunlight which are not so hot in	Bedroom is used for sleeping; hence temperature should be cool here. Evening

			nature should enter in.	sunrays which are cool in nature, enters in and takes out dampness and germs, and arouse sleeping habitat.
3	Hall/ Living room/ Drawing room	N/NE	Sunlight is available utmost time of the day	As occupants utilizes it for maximum time in day, so more light should be available here.
4	Study room	N	More sunlight	This room should be well lighted for studying purpose and should not stimulate sleeping environment.
5	Store room	N	More sunlight	More light enables us to find stored things quickly and also shields us from dashing our leg or other parts

				into stored things as this room is messed around already.

(N-North, S-South, E-East, W-West, NE-Northeast, NW-Northwest, SE-Southeast, SW- Southwest).

The aspects are designated on the thrust of Sun-path diagram given below.

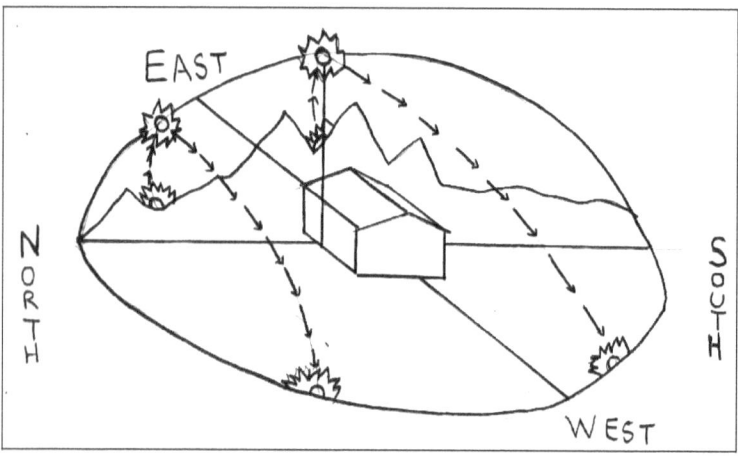

The other principles of planning are mentioned below:

2] Prospect:

It's about the outside view we get from inside of the house by proper projecting the windows and doors.

a) **Living Room:** Prospect of living room i.e. hall, should be in connection with main road to keep control on the plot.
b) **Bedroom:** It is generally on the rear side of the house to avoid disturbance of main road, traffic and noises.
c) **WC (Water closet) & Bathroom:** Water closet and bathroom should also be on the back side of the house, as the pipe lines ruins the aesthetic look of the house. If by any fluke these units are positioned on the forefront of the house then make sure you provide duct for pipelines.

3] Privacy:

It is a state in which one is not observed or disturbed by other people. Privacy can be inter houses or inter rooms.

The privacy of the house can be achieved by planting moderate trees, building compound walls and even by providing windows at higher level than the average human height.

Privacy of one room from other can be achieved by providing door on one side of the longer wall, as the person standing in the door can only see the minimum portion of the room.

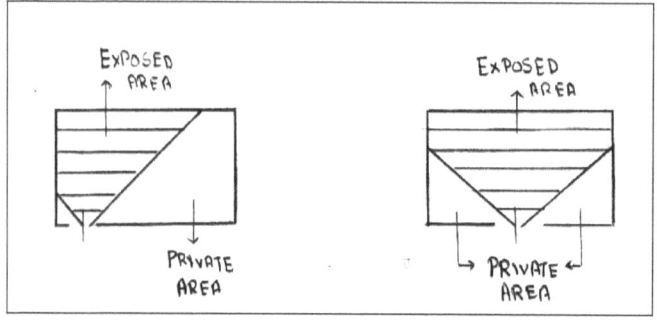

4] Roominess:

It is the feel of speciousness obtained in the room. It is related with the dimensions of the room, i.e. room's length, width and height. The ratio of length and width should be so regulated that it should not furnish with a tunnel or railway bogie like feel. Square rooms always appear smaller than rectangular rooms of same area.

5] Grouping:

Grouping means arranging rooms sequentially with respect to their functions and utility.

1. Verandah should be the first unit of the house.
2. Hall / living room should be next to verandah.
3. Kitchen and dining should be contiguous to each other.
4. Store room and kitchen should be adjusted.

5. Sanitation (WC & bathroom) should be adjacent to bedroom.
6. Staircase should be approachable from each room.

6] Circulation:

Circulation means movement of occupants in vertical and horizontal direction.

1. <u>Vertical circulation</u>: It is the movement in upward or downward direction. It is generally achieved by stair-cases. Other examples are ramps, lifts, etc.
2. <u>Horizontal circulation</u>: it helps in movement in different units of building on same floors. Example – passages, corridors, lobby, etc.

The circulation should occupy 15% of overall built up area.

7] Sanitation:

There should be proper sanitary units in ones house. The WC and bath should be so adjusted that there should be satisfactory ventilation and light.

They should be located close by bedrooms, as during midnight pee we do not have to walk all the way to the other corner of the house for peeing and other purposes.

8] Elegance:

Elegance is the look of the house we come to have from outside. The house should be so planned that it gives a delightful look from outside by increasing plinth height, providing arches, planning rooms at different heights, using decorative stones, combination of flat and sloping roofs etc.

Elegance depends upon planning as well as elevation. Without elegance, a properly planned building may not look alluring.

9] Furniture Requirement:

While planning, the furniture requirement of the owner should be known. It helps in determining that there is enough space remaining for movement in the room after placing the furniture.

- Minimum size of bedroom should be such that it can accommodate a cupboard, a dressing table and a double size bed. Other shrimpy requirements are bookshelf, chair, table, etc.

- Living room should be capable of holding a sofa set, teapoy, T.V. Cabinet, shoe rack, etc. and still have enough space for movement.

- Dining room size is decided upon the number of dwellings.

Example- if 4 members are living then a 4 people dining table should fit in there.

10] Parking Area:

Parking spaces for cars, scooters and bicycles shall be provided.

The space requirements for each are given below:

Car – 2.5 x 5 meter

Scooter – 3x3 meter

Bicycle – 1.4x1.4 meter

11] Economy:

The factor which may limit all the above principles is the economy. All efforts are in vain if the cost is not considered. In any case aspect and circulation should not be disturbed for economy. On other hand, prospect can be ignored over aspect. Economy can be achieved by simple building plans along with minimum doors and windows and by using various construction management techniques. In short planning should be done chewing over the budget of the house owner.

❖ Limiting Dimensions

- **Habitable rooms –**

    Height – minimum 2.75 meter.

    Size – minimum 9.5 square meter.

    Width – minimum 2.4 meter.

- **Kitchen –**

    Height – minimum 2.75 meter.

    Size – minimum 5.5 square meter.

    Width – minimum 1.8 meter.

- **Kitchen with dining –**

    Height – minimum 2.75 meter.

    Size – minimum 9.5 square meter.

    Width – minimum 2.4 meter.

- **Bathroom –**

    Height – minimum 2.2 meter.

    Size – minimum 1.8 square meter.

    Width – minimum 1.2 meter.

- **Water Closet (WC) –**

    Height – minimum 2.2 meter.

    Size – minimum 1.1 square meter.

    Width – minimum 0.9 meter.

- **Combined Bath and WC –**

    Height – minimum 2.2 meter.

    Size – minimum 2.8 square meter.

    Width – minimum 1.2 meter.

- **Store room –**

    Height – minimum 2.2 meter.

    Size – minimum 3 square meter.

a) **Staircase –**

    Width – minimum 0.75 meter.

    Thread – 300 mm.

    Riser – 150 mm.

The above all principles are made to achieve the Bye - laws of the building.

❖ Building Bye – Laws as it is:

- To curb the haphazard growth of towns or cities.
- To facilitate future use of land, widening of streets, controlling the ribbon development in an area.
- To reduce pollution in area by restricted population density in an area there by providing hygienic environment.

- To ensure that every citizen will receive facilities like water supply, sanitation, ventilation, electricity supply, parking and safety.
- To help architects, engineers in planning and design of every civil engineering construction.
- To ensure that every citizen will safe against fire, pollution, health hazard and building failure.

❖ Open spaces around the house

It is critical to assess the spaces around the house in planning and designing of house, otherwise you have to compromise with the left over spaces.

Giving notion to the spaces around house rather than depending on the left over is a good planning practice.

Every house should keep the following –

1. Front open space.
2. Rear open space.
3. Side open space.

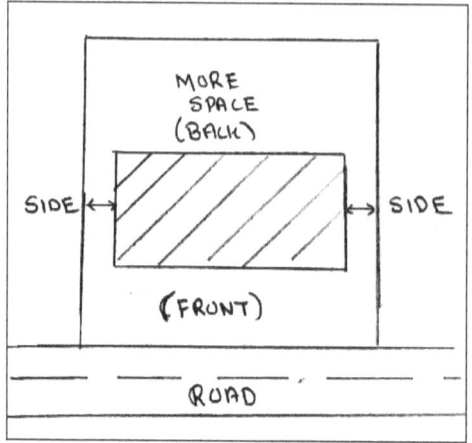

1) Front Open Space –

Front yard is unfailingly connected to street, so it should be allotted for non – private works like parking, gardening, pet house, driveway, etc.

For front yard the average width should be 3 meter.

2) Rear Open Space –

Back yards are isolated and away from roads hence more private tasks are carried out here. Tasks like pet bathing, barbeque cooking, open sky dinner with family, tenting, etc. the above all are not common in India but still I have mentioned it. (In India back yards are used for drying washed clothes, storing old furniture, etc.) .

Swimming pools are advisable to be constructed in the rear space. Playing areas for kids should be in back yard as it is away from road and traffic, hence safe for kids.

Also septic tank should be located in the back yard. The average width of the back yard is 4.5 meter; it should not be less than 3 meter. More than 4.5 meter is allowable according to the plot area. It has more area compared to front and side open spaces.

3) Side Open Space –

It is the space given for ventilation of air. It should maintain a proper privacy from neighboring houses. Sufficient space should be available to move from front yard to back yard. The width should not be less than 1meter, or if by any chance the side space is facing the lateral road then its width should not be less than 3 meter.

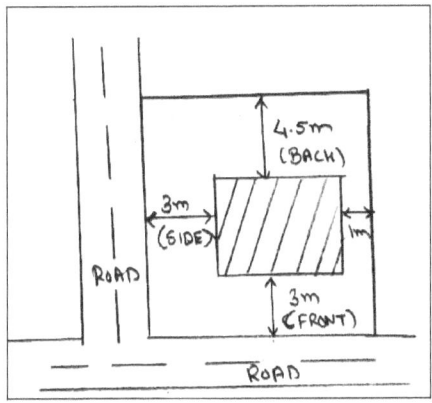

The front open space also depends upon the width of the road, where as the side and rear open space depends upon the height of the house.

Width of the street	Front open space
Up to 7.5m	1.5m
7.5 – 18 m	3m
18 – 30 m	4.5m
Above 30 m	6.0m

Height of House	Left around space
10m	4.5m
15m	6m
18m	7m

❖ Height Of the House

It is necessary to figure out and control the overall height of the house so that it does not overcast its shadow on roads and other adjoining areas.

If rules were not made regarding height of the house then every individual would build their house up to the desired height and all roads would be overcastted by shadow and would become sunless, which is not acceptable as roads should always be illuminated by sunlight in daytime.

Certain rules are made by the authorities which should be followed. The height of house depends upon 2 factors.
1. Width of the street in front.
2. Width of the rear space.

1) Width of the street in front –

The height of house depending on the street width is given in the following table.

Width of street	Height of the house
Up to 8m	Maximum 1 ½ times width of street. i.e. if the width is 6m then height should be 6x1 ½ i.e. 6+3 = 9 (3 floors)

8 to 12m	Not more than 12m (4 floors).
Above 12m	Between the width of the street and not more than 21m (7 floors)

2) Width of rear space –

The height of house with respect to rear space is fixed by imaginary diagonal line which makes an angle of 63 ½° with the horizontal at the point of rear boundary.

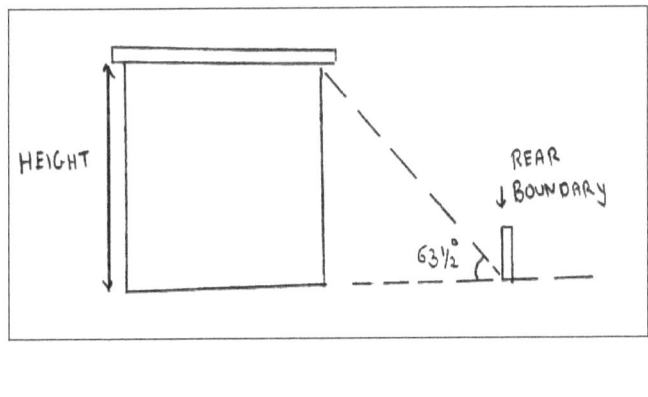

❖ What if planning bye laws is not done?

If everyone makes buildings without obeying any rules then there will be problems of –

1. Irregular and narrow roads.
2. Frequent traffic.
3. Problems of parking.
4. Health problems due to pollution.
5. Poor light and ventilation.
6. Problems regarding services like water supply, drainage, electricity, gas, etc.
7. Noise, if not build in residential zone.

After plan is finalized the cost estimate, developed plan, center line plan, site plan, elevation and foundation plan are prepared.

The sense of plans are given below –

Developed plan – it is the house plan showing the arrangement of rooms and thickness of walls from top view.

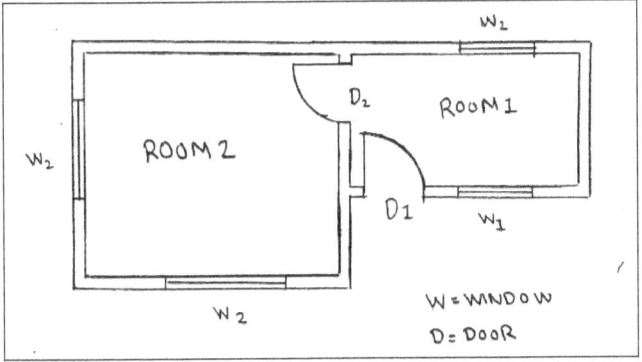

Center line plan – shows line passing through the center of the walls.

Foundation plan – shows the location, arrangement and size of footings.

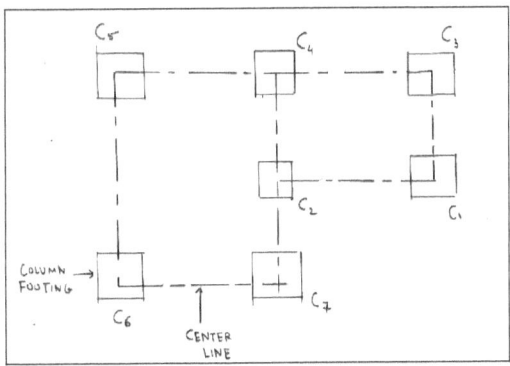

Site plan – It shows the complete property and identifying all structures in relation to property boundaries.

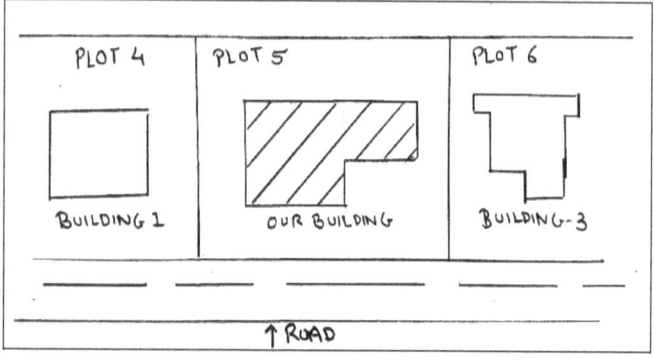

Elevation – it shows exterior view of the house.

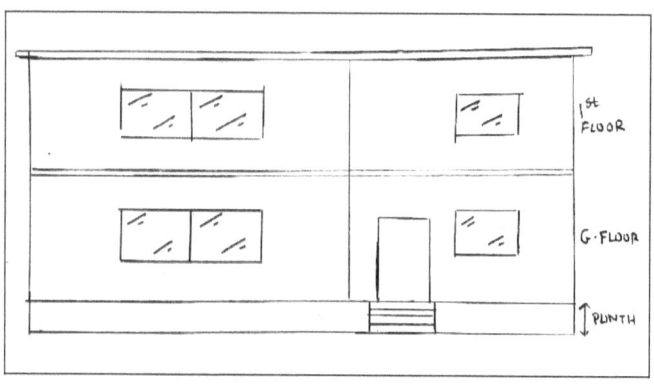

Chapter 4 – Necessary Paper Works

Now that we have planned the house, it's time for getting permit for building the house.

If the house is raised without permission then it is treated as illegal and that can be demolished at any instance without prior notice by the capable authority.

---

1) What is a building permit?

A building permit is a legal evidence of permission one gets to begin the construction.

Once you hold the building permit then your construction would be legal and no one has the liberty to cease or flatten the construction work in future.

Secondly, building permit is about safety. By enforcing construction standards, it gives you and other occupants of the house the best chance to avoid fire, structural failure or something as simple as hitting your head on the door frame while entering a room.

2) Who grants the building permit?

The capable authorities are according to the sum and substance of site whether in village, town or in a city.

These authorities are-

1. Grampanchayat.
2. Municipal Corporation.
3. Collector of district.
4. Town planning authority.
5. Development authority.

3) Documents required for getting permit.

1. Building plan.
2. Site plan.
3. Recent 7/12 extract.
4. Document of ownership of land.
5. City survey plan.
6. Affidavit on rs 100/- stamp paper.

7. (If area of plot is <500 m²)

8. Noc from C.A.U.L.C (if plot area is > 500 m²).

Post Script – For seeking permission of construction work, its drawing drawn to a scale of 1:100 (i.e. 1 meter on land is 1cm on paper) are submitted which includes building plans, elevation, section, detail drawings, site plans, area statement (detail of plot area, built-up area, plinth area, carpet area, etc.), schedule of opening (dimension of doors and windows), with signature of architect and owner. These drawings are prepared by keeping rules and regulations in mind.

4) What happens inside?

The capable authority examines the paper and plan submitted. The authority approves the drawings submitted if drawn as per bye laws; otherwise the owner is informed to submit the corrected drawing before the specified date to seek the permission.

5) How much time it takes?

It depends on the complexity of the project. For simple buildings it takes about 2 weeks and for large or complex buildings it takes 8 weeks.

The sanctioned plan remains valid up to 3 years but recently it has been dropped down to 1 year.

6) Cost of permit?

Fees are based on the value and total floor area of the building. It is required to pay 75% of the fee while submitting the documents and rest 25% when you pick up the permit. The validity of permit can be extended by paying some extra charge before expiry of the permit.

CONCRETE AND STEEL

❖ Concrete -

Grade of concrete	Mix proportion cement: fine aggregate: coarse aggregate
M 7.5	1:4:8

M10	1:3:6
M15	1:2:4
M20	1:1.5:3
M25	1:1:2
M30	1:1:1

The grade designates M for 'mix' design of concrete followed by the compressive strength number in N/mm^2

E.g. – M20 The concrete made of grade M20 would have compressive strength of 20 N/mm^2 in 28 days. I.e. 2.04kg/mm^2

Which means a small surface area of 1mm^2 can bear a load of 2.04kg.

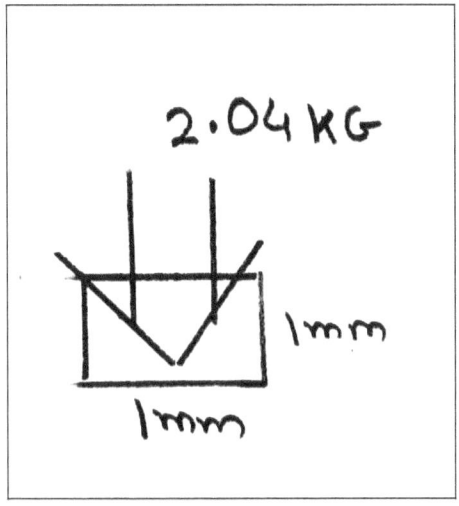

The respective mix proportions indicates ratio of cement: fine aggregate: coarse aggregate

I.e. cement: sand: khadi(kankar)

E.g. – M10 ⟶ 1:3:6

This means 1part of cement, 3 parts of fine aggregate (sand), 6 parts of coarse aggregate (khadi) are taken to make unit quantity of concrete.

FOR 1 M³ OF CONCRETE

SR NO	Proportion	Cement (in bags) each 50 kg	Sand (in m³)	CA (in m³)
1	1:4:8	4	0.47	0.94
2	1:1.5:3	8	0.405	0.81
3	1:2:4	7	0.44	0.88
4	1:3:6	5	0.45	0.91
5	1:5:10	3	0.475	0.95

❖ Steel

Grade	Yield strength
Mild steel (Fe250)	250 N/mm² = 25.49 kg/mm²
High yield strength deformed bars (HYSD TOR40) (Fe 415)	415 N/mm² = 42.32 kg/mm²
TOR steel (TOR 50) (Fe 500)	500 N/mm² = 50.99 kg/mm²

Percentage of steel required in following works

Works	% of Steel
Foundation	0.5 – 1%
Column	1 – 3%
Beam/lintel	1.5 – 3%
slab	0.9 – 1.5%
Shear wall	1 – 2%

Chapter 5 – Foundation

❖ Types of structures –

a) Load bearing :
- In load bearing structure, each wall is structural member and weight of slab is carried by the walls and transferred to the foundation.
- This is an outdated structure and is not into practice nowadays.

b) Framed structure :
- In framed structure wall do not carry the weight of the slab.
- The structural units are beams and columns.
- The walls are only for partition. The weight of wall is carried by beam under it.
- The weight of the slab is carried by beams. Beams transfer the self weight and load of slab to the column. Further column transfers it to the foundation and foundation transfers the overall load over a greater area to the below soil.

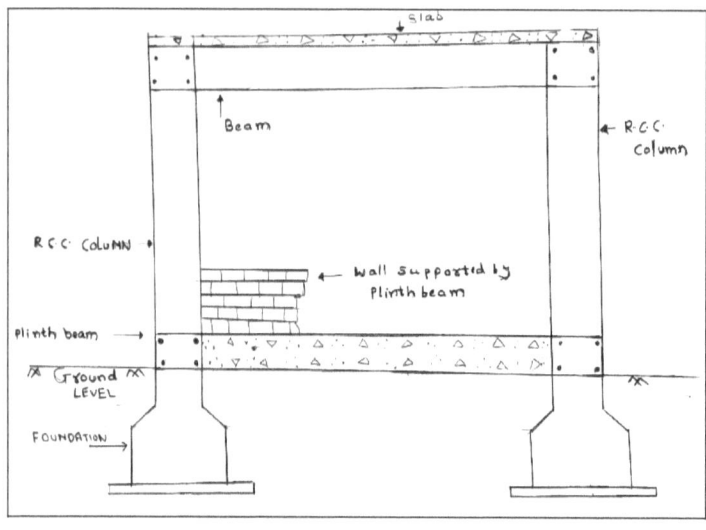

❖ Foundation –

Foundation is a structure which is not visible as it is under the ground. On that account it is also termed as substructure. It is the important fraction of the house as it supports the overlying structure (super structure).

Foundation do not typically contribute to the architectural aesthetics of a house, yet without suitable foundation a house will not function effectively, will be unsafe and its architectural merits will rapidly fade. The main function of the foundation is to take load from the superstructure via column and divide out evenly over a larger area to the soil strata underneath it.

❖ Design of foundation –

The size of the foundation is determined by the overlying load of the house. (The calculation for the same is assumed in this book and not calculated as mentioned in the disclaimer).

Factors affecting design of foundation –

- Soil type and water content.
- Site condition / Environment.
- Structural requirements.
- Construction requirements.
- Economy.

Foundation should be so designed that it should withstand the load fallen on it without any settlement. A small settlement is allowable within the tolerable limits, Also while designing the effective cost cutting element should be kept in mind.

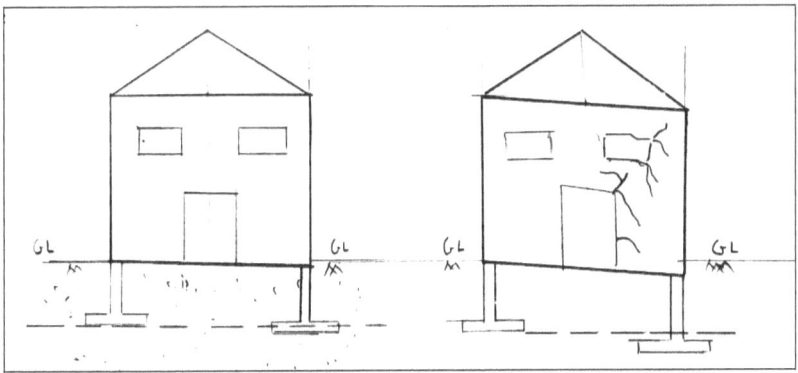

In simple words, consider 1 cubic meter of concrete; it weighs averagely about 2400 kg. Now just think how much m³ of concrete stand in need to construct a 2 storey (floor) house, and also the quantity of bars needed.

Hence, the weight of a two storey house is a huge number, and foundation should be strong enough to bear and transfer that amount of load over a large area on to the subsoil.

Foundation does not limit itself to only load bearing ability, addition to that it possess many other tasks like reduction of load intensity, safety against sliding, safety against undermining, provision to level surface, even distribution of load, protection against soil movements, etc.

1) Safety against sliding & overturning –

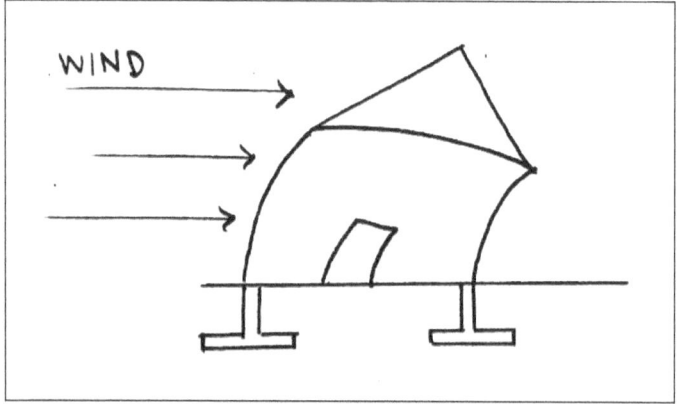

Foundation anchors the house to the ground and hold out against the horizontal forces like wind, earthquake, tsunami, etc, hence increasing the stability and giving a wide berth to overturning.

2) Reduction of load intensity –

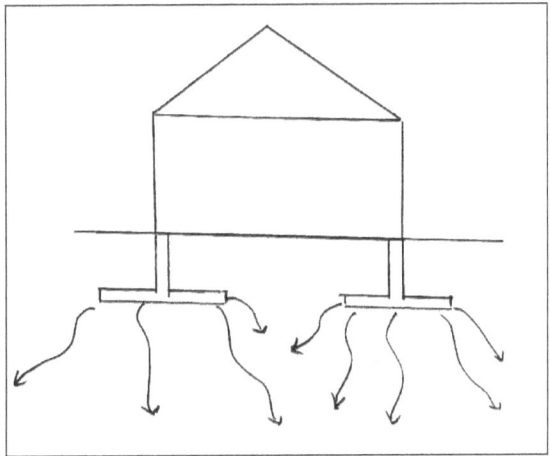

It acts as a middleman between the superstructure and the subsoil and spreads the load of the superstructure over a larger area to the subsoil, so that the load intensity is brought down and do not run over the safe support capacity of the soil.

3) Provision of level surface –

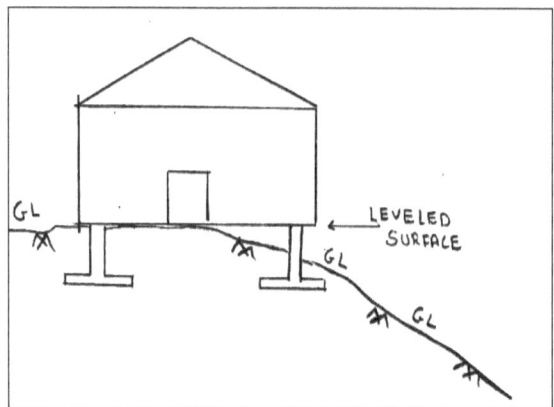

Imagine building a house on a steep slope, the slope will not permit you and also will not provide you a leveled hard surface to build a house on. Here foundation makes it possible by compensation the mess and providing leveled and hard surface over which the super structure can be built.

4) Safety against undermining –

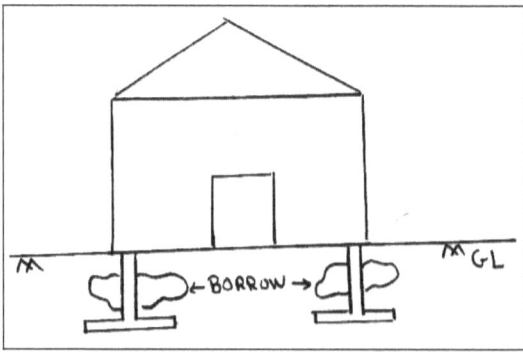

It contributes to the safety against undermining or scouring caused due to small burrowing animals like mole, rats, snakes and also due to water.

5) Even distribution of load –

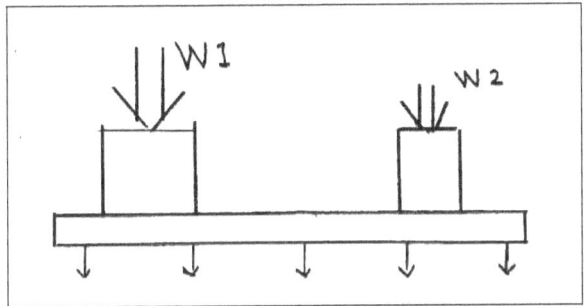

When uneven loads are accumulated at a single point then the foundation distributes this non-uniform load evenly to the subsoil.

Eg. – 2 nearby columns carrying uneven load can have a combine footing which would transfer the load to the subsoil evenly.

6) Protection against soil movement –

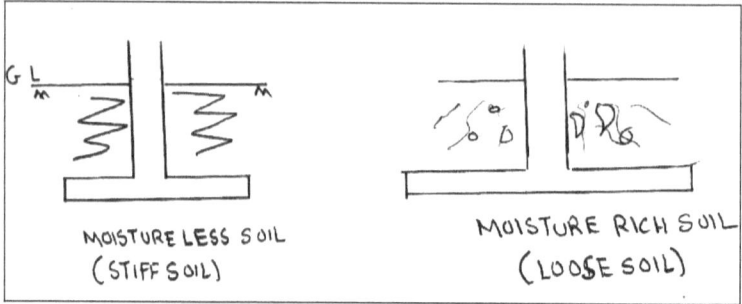

Foundation prevents or minimizes the distress in superstructure due to expansion or contraction of the subsoil because of moisture movement of soil.

I.e. when the soil is moisture deficient it gets contracted and tightens up around the footing. And when the soil is moisture rich, it get loosen up around the footing. But in both cases the superstructure is unharmed and unmoved.

❖ Types of foundation

The classification of foundation is clinch on basis of its depth and width relation.

The two extensive types are shallow foundation is less than or equal to its width than it breaks under **shallow foundation** $(D \leq B)$.

And in case of deep foundation the depth of the foundation is more than its width $(D>B)$.

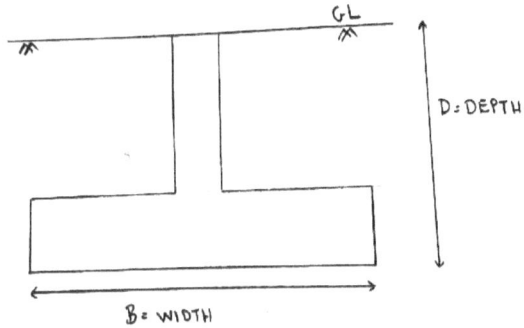

- ## **Shallow Foundation**

A shallow foundation transfers the load to the subsoil which is in the vicinity of the surface of the ground. I.e. a shallow foundation is used when there is a hard soil stratum which can bear the load of the house, is available at a short depth. Shallow foundation are customarily used in house construction as a 2 storey bungalow load is bearable by soil at shorter depth as compared to high rise buildings, whose load bearability requires deep foundation.

The extensive shallow foundation system further fragments into two commonly used types –

1. Isolated foundation
2. Raft foundation

1) Isolated Foundation –

These type of foundation are used when the underlying soil has a load bearing capacity more than

$24KN/m^3$ i.e. 2447 kg/ m^3 i.e. a meter cube of soil must be able to bear at least a load of 2447 kilograms.

Here in a separate footing bed is constructed for individual columns. They are generally square, rectangular or circular in section. This foundation type is most economical, requires less labour expertise and has ease of constructability as only portion where the footing has to be placed is excavated.

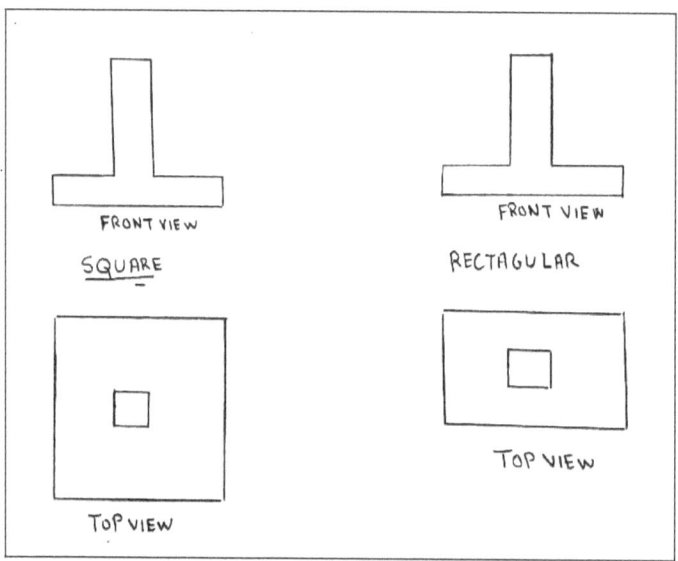

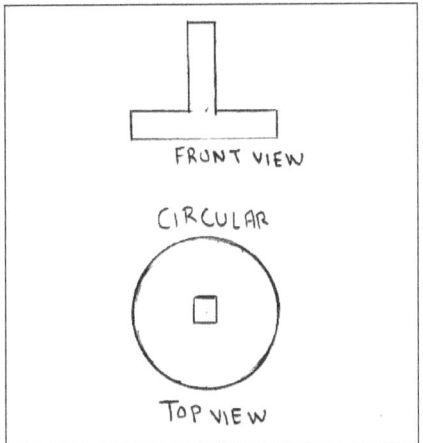

Further the isolated foundation are classified as –

1. **Flat or Pad footing –**

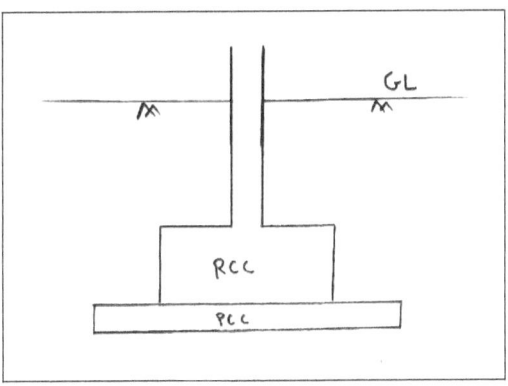

It has a layer of plain cement concrete without reinforcements (steel bars). Over which the reinforced cement concrete footing with flat surface is constructed.

2. Stepped footing –

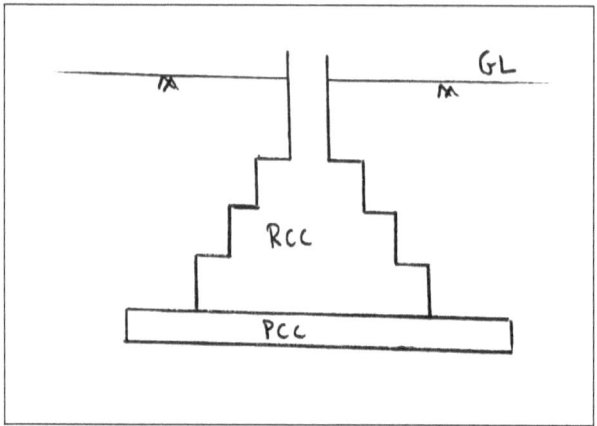

These footings are called stepped because of the way it appears. These were used in old days and at present they are not generally in practice.

3. Sloped footing / Trapezoidal footing –

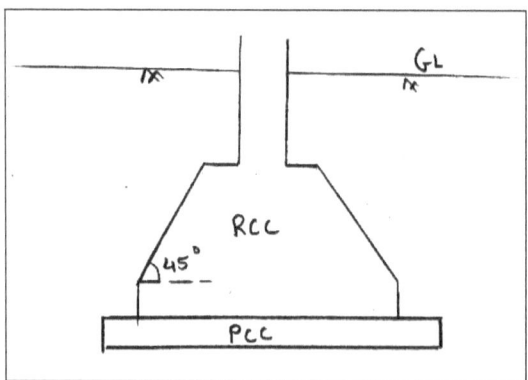

These footings are the most commonly used footing these days. It is called as trapezoidal footing due to its shape. The concrete used as compared to flat or stepped footing is less, hence economical.

2) Raft Foundation –

If the underlying soil has a load bearing capacity less than 24KN/m^3 i.e. 2447 kg/ m^3 i.e. a meter cube of soil is not able to bear at least a load of 2447 kilograms; then raft footing comes into play.

They are provided either when the load is too heavy or when the safe bearing capability of soil is less. These are constructed in areas where the water table of the ground is at a higher level. For eg – Coastal areas.

In raft footing the whole built-up area is excavated and a concrete reinforced mat or bed is laid so that it will spread the load over larger area with less depth. Then the columns are constructed on the constructed mat at the specified position.

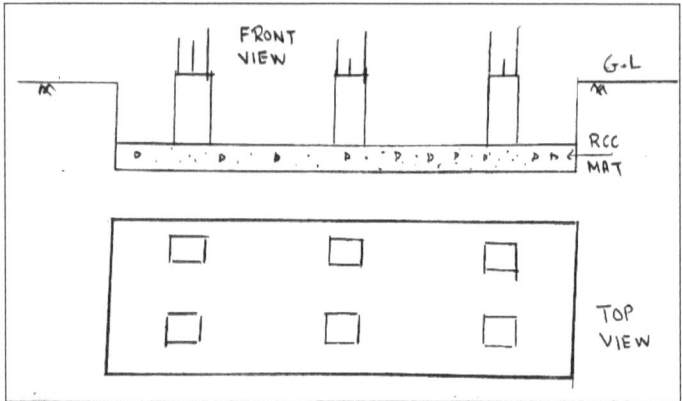

This is very uneconomical system, but safety comes first.

• Deep Foundation

Deep foundation transfers the load to the subsoil which is deep down the ground surface. i.e. a deep foundation is used when there is a hard soil stratum which can bear the load of the house, is available at a deeper depth.

Deep foundations are used for high rise buildings or other buildings whose load is too heavy to bear by the soil. This type of foundation is not generally used for 2 storey house construction.

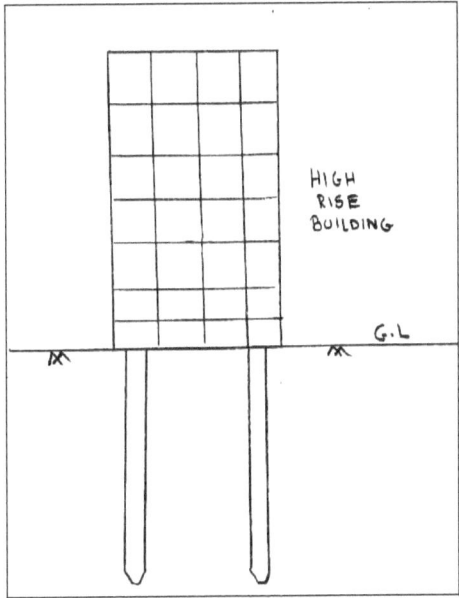

Types of deep foundation

1. Pile foundation.

2. Well foundation.

3. Caissons.

4. Cofferdams.

❖ Construction procedure –

1. Preparing Sheds –

Before starting the procedure it is called for safeguarding the materials that will be required for construction. For that at least a temporary shed is needed to be assemble which should be rain coated. This provision becomes mandatory in monsoon season.

Materials like cement and steel bars should be stored with care, because if cement comes in contact with water it begins to settle and becomes hard, and steel bar catches corrosion.

2. Marking layout on ground –

Marking is done so as to get the true outline of the house and the area to be excavated. Marking is done with white lime powder. There are two methods for lay outing, namely

a) Center line method

b) Face line method

A. Center line method –

- In this method the layout is fixed by fixing the center line of the foundation plan

- For this purpose, one corner of the foundation is fixed by measuring the distance from the border of the plot and a wooden peg with a nail on top is driven to mark the corner.

- Then the string is tied to the nail on the top of the peg and the center line is marked on the ground using the string as per the foundation plan.

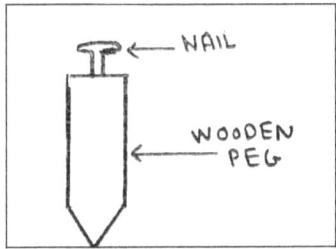

- Then according to footing size of each column, markings are made around center of each column.

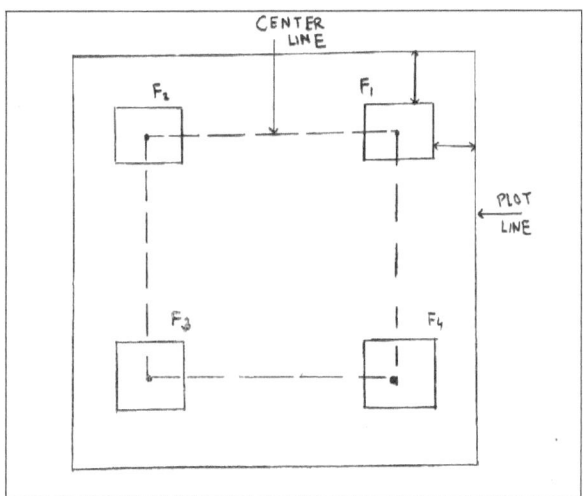

Hence we got the center line of the house and the quadrilateral areas to be excavated, on the ground.

345 Rule (Katcone)

This rule is used to square up the lines marked. I.e. to determine each line is at a right angle with other line. This is necessary to maintain the shape of the house and eliminate unevenness of the house.

- Pythagoras triplets are used i.e. 3, 4, 5.

- If a triangle is made using 3 lines measuring 3, 4 and 5, it will construct a right angle triangle.

- For this a base line is marked using the measurements in the given plan.

- In the given figure below, let the base line be 'line 1' and let 'line 2' be the other line which is needed to square up with line 1.

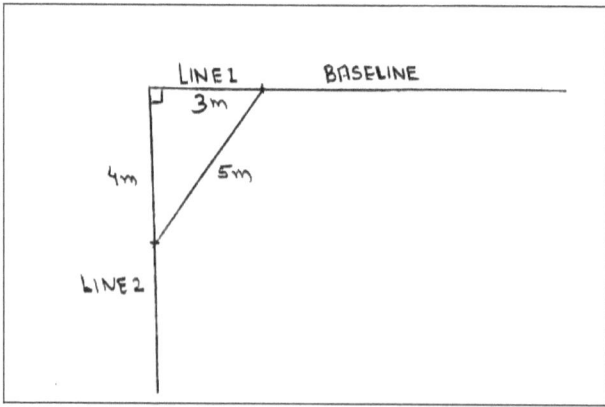

---Start To End On How To Build A House---

- Now mark a point at 3m on line1 and 4m on line2, as shown in figure.

- Then the angle between two lines should be so adjusted that the distance between two marked points is 5m.

- Once the distance between them is fixed at 5m then the angle obtained is right angle and the 2 lines are squared up.

- This procedure is repeated until the layout is completed.

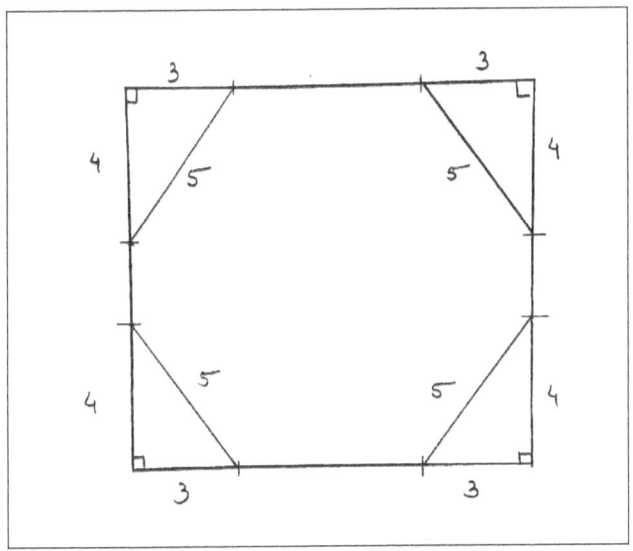

- The Pythagoras triplets can be increased if the line measurements are big. Bigger triplets are multiples of 3, 4, 5 i.e. (6, 8, 10 or 9, 12, 15...).

- At last the diagonals of the quadrilateral formed are measured, if they measures equal then the lines are rightly squared up or else there is an error in the process.

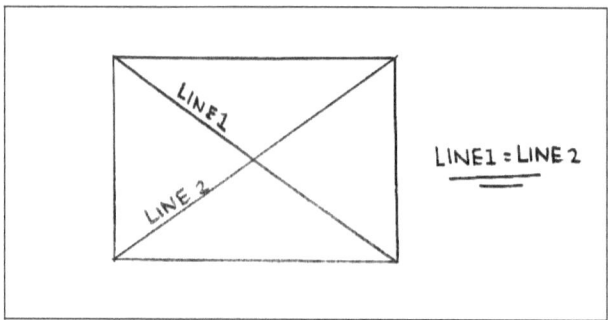

Center line method is chiefly used as it is an accurate method and margin for errors are less as it is hinged on Pythagoras principle.

B. Face line method –

Here the center line is not taken into account. Each face of the column footing is measured from the edge of the plot and then it is marked on ground.

This method is not used mostly as it is less accurate and more chances are there for errors.

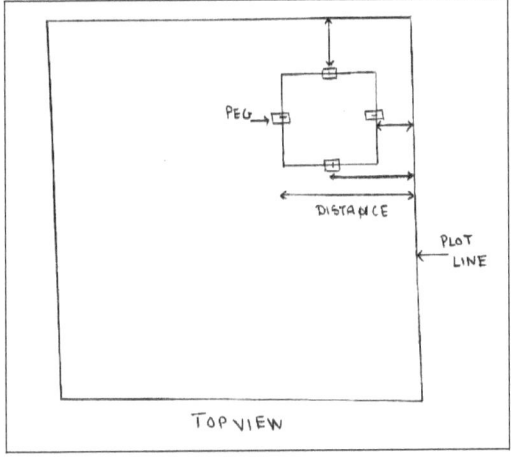

3. Excavation –

- Pits (trench) are dug from the area which is marked.

- It is usually done by hand tools in small house construction.

- Sides of the trenches shall be vertical and its bottom shall be perfectly leveled, both longitudinally and transversely.

- During excavation if rocks are found those shall be leveled as far as possible and the small spaces which are difficult to level shall be filled with concrete.

- No material excavated from foundation trenches shall be placed nearer than one meter to the outer edge of the excavation.

- Timber wood supports are given in case of soft soil as it may collapse.

- If the soil consists of hard murum, it does not need any brace after the excavation of trench.

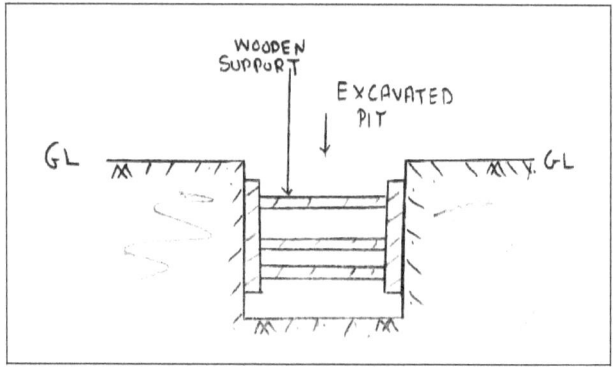

NOTE – While excavating, the area for septic tank is also dug.

4. Raising structure –

Step1 - Spread a layer of PCC (PCC do not have steel bars) i.e. plain cement concrete of proportion 1:4:8 or as fixed, over the surface area of excavated pit up to the calculated height. This layer provides a leveled surface for footing.

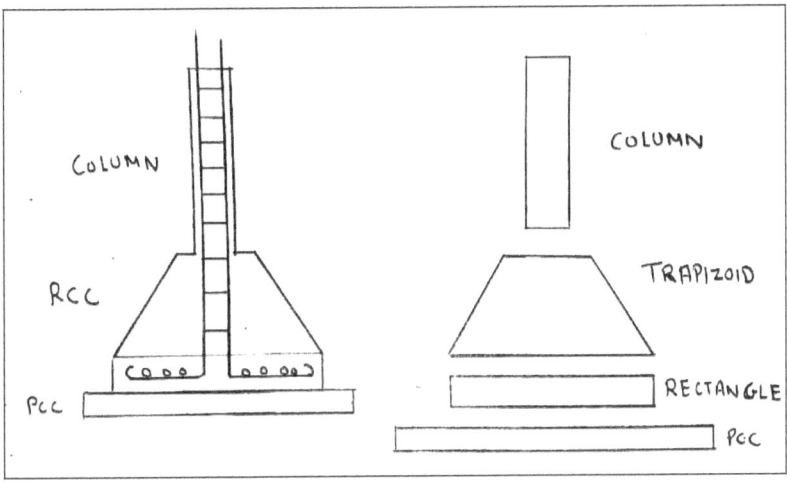

Step2 – The center point is marked on the PCC bed. This is done by coinciding the center line in X and Y direction and then transferring the point by using plum bob.

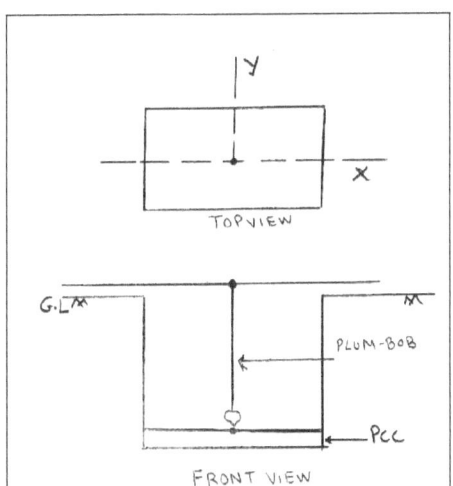

Step3 – Prepare a grid of steel bars of needed expanse. This is done by cutting reinforcements i.e. steel bars of desired length and are tied together using binding wires. This grid is placed at the center on the PCC bed. The grid is actually placed on cover blocks so that the steel bars are totally covered into the concrete and do not approaches the finish of the concrete surface, as it may cause rusting of steel.

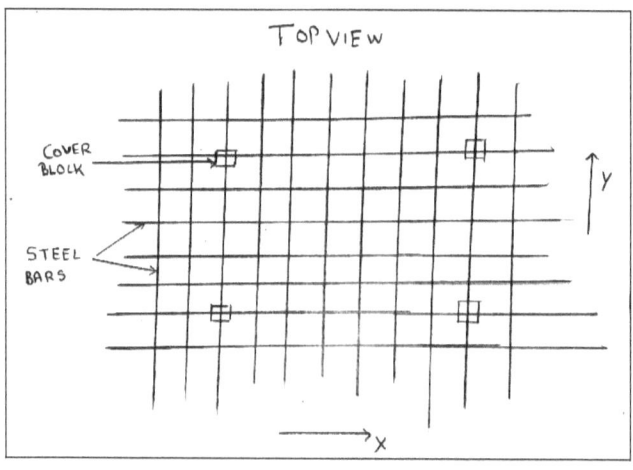

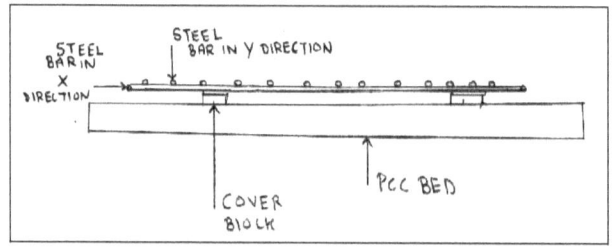

(Note – binding wires are thin metal wire use to tie steel bars)

Step4 – Preparing the frame work of column. The detail process is discussed in column chapter.

The frame work of column is placed over the grid at the center. The verticality of the column structure is checked using plum bob.

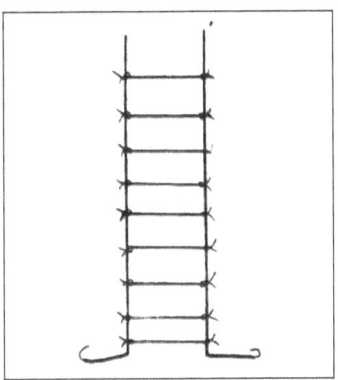

Step5 – Construction of rectangular and trapezoidal part. A formwork (mold) of wood is prepared for rectangular part in which the concrete of mix ratio 1:2:4 or as fixed, is poured and is vibrated using mechanical vibrator or by tamping rod. Then more concrete is poured on it and the shape of trapezoid is formed using trowels and other hand tools. After a day the formwork is removed and the concrete is watered.

Note – Vibration is necessary as it removes the entrapped air in the concrete and helps in consolidating the concrete.

Step6 – Next is construction of column. Wooden formwork for column is prepared of dimensions using design data. This formwork is made to enclose the steel framework which is already jammed into the rectangular and trapezoidal concrete. Wet concrete is poured into the formwork and is vibrated. The formwork is removed after 24 hours when the concrete is hard and settled.

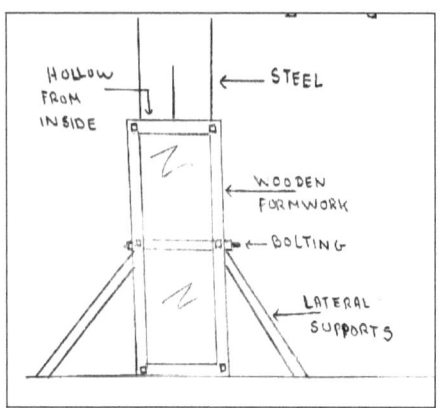

This above is for only one column footing, all other column footings are constructed similarly and simultaneously.

The whole foundation structure is watered regularly.

Then after, the remaining portion of the pit is filled with soil from the near land and letting only small part of the column and it's steel to remain above the ground.

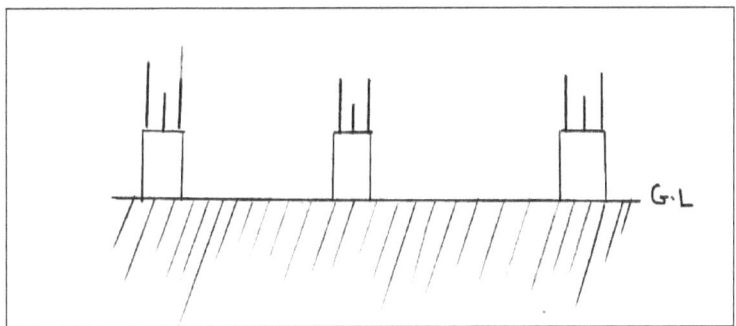

- **Precautions –**

1. Check the excavation for its dimensional accuracy before laying foundation.

2. Secure the vertical rods in foundation firmly in its position.

3. Ensure proper compaction of concrete in the sloped portion of foundation.

❖ Material calculation
1. PCC bed in foundation –

For plain cement concrete bed (PCC) in foundation the best proportion used is 1:4:8 i.e. M7.5 grade concrete.

Consider 1 cubic meter of volume of work to be done.

∴ Wet volume = $1 m^3$

In order to convert dry volume from wet volume, the quantity should approximately increase by 33%.

∴ Dry volume = 10 + (10 × 33%)

$$= 1.33 \ m^3$$

Consider wastage of material as 15%

∴ **Final volume** = 1.33 + (1.33 × 15%)

$$= \mathbf{1.529 \ m^3}.$$

Now for PCC work materials needed are –

Cement,

Fine aggregate (sand),

Coarse aggregate (Khadi / kankar).

Quantity of cement = (Final volume / Addition of ratio) × Part of cement

$$= (1.529/\ 1+4+8) \times 1$$

$$= 0.1176 \text{ m}^3.$$

Quantity of Sand = (Final volume / Addition of ratio)

$$\times \text{ Part of sand}$$

$$= (1.529/\ 1+4+8) \times 4$$

$$= 0.47 \text{ m}^3.$$

Quantity of coarse

   aggregate = (Final volume / Addition of ratio)

$$\times \text{ Part of CA}$$

$$= (1.529/\ 1+4+8) \times 8$$

$$= 0.94 \text{ m}^3.$$

Water cement ratio for PCC is 0.35

∴ Water/ cement = 0.35

∴ Water/ 0.1176 =0.35

∴ Water =0.04116

∴ **Water = 41.16 litre**

Now, 1 bag of cement = 50kgs = 35litre =0.035 m³.

Cement quantity in cum. = 0.1176

Cement quantity in no. of bag = 0.1176/0.035

$$= 3.52 \text{ bags}$$

$$\cong 4 \text{ bags}$$

So for 1 m³ Of PCC work it will require 0.1176 m³ of cement i.e. 4 Bags, 0.47 m³ of fine aggregate (sand), 0.94 m³ coarse aggregate (khadi) and 41.16 litre of water including wastage.

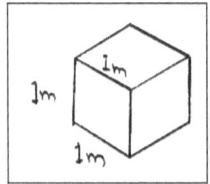

2. RCC in foundation –

For reinforced cement concrete (RCC) the best ratio used is 1:2:4 i.e. M15 grade concrete.

For foundation the steel % required is 0.5% – 1%.

Consider 5 m³ volume of concrete work required and 1% steel.

Wet volume = 5 m³.

Increase by 33% for dry volume

∴ Dry volume = 5 + (5 × 33%)

$$= 6.65 \text{ m}^3.$$

Increase by 15% for wastage

∴ **Final volume** = 6.65 + (6.65 ×15%)

$$= 7.65 \text{ m}^3.$$

Materials required for RCC work –

Cement

Sand

Coarse aggregate

Reinforcement (steel bars)

Quantity of cement = (Final volume / Addition of ratio)

$\times$ Part of cement

= (7.65/ 1+2+4) × 1

= **1.0928 m³.**

= 1.0928/0.035

= 31.22

$\cong$ **32 bags**

Quantity of Sand = (Final volume / Addition of ratio)

× Part of sand

$$= (7.65/\ 1+2+4) \times 2$$

$$= 2.185 \text{ m}^3.$$

Quantity of coarse

aggregate = (Final volume / Addition of ratio)

$$\times \text{ Part of CA}$$

$$= (7.65/\ 1+2+4) \times 4$$

$$= 4.37 \text{ m}^3.$$

Consider water cement ratio to be 0.45

Water/Cement = 0.45

Water/1.0928 = 0.45

Water = 0.45 × 1.0928

Water = 0.4917 m^3.

Water = 491.7 litre

Reinforcement = 1% of wet volume

$$= 1\% \times 5$$

$$= 0.05 \text{ m}^3.$$

Density of Reinforcement = 7850 kg/m³

∴ **Quantity of reinforcement** = 0.05 × 7850 kg/m³

= 392.5 kg

= 0.392 tonnes

=3.925 quintals

Quantity of binding wires = 1 kg of binding wire is required per quintal of reinforcement.

(1 quintal = 100kg)

∴ **Quantity of binding wires = 3.925 kg**

∴ For 5 m³ of RCC work materials required are –

32 bags of cement (each 50kg)

2.1857 m³ sand (fine aggregate)

4.37 m³ coarse aggregate (khadi)

491.7 litre water

392.5 kg steel bars (reinforcements)

3.925 kg binding wires

━━━━━━━━━━━━━━━━

Chapter 6 – Plinth Beam

The beams which are constructed at ground level are called as plinth beam. They are components laid horizontally across the periphery of the house above the foundation.

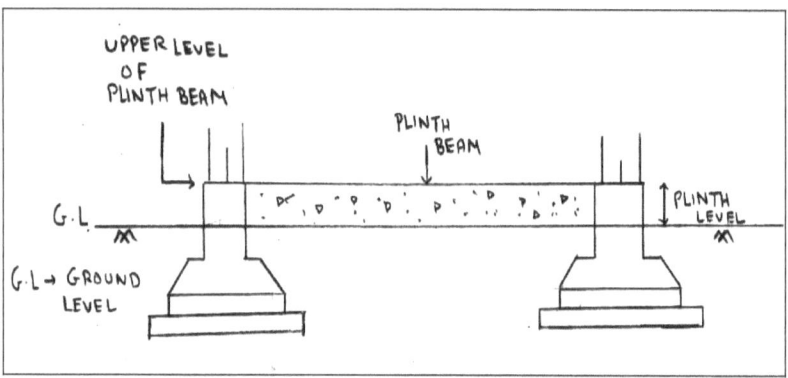

The main purpose of plinth beam is to support the walls which are built over it. In frame structure the walls do not carry any load, but the brick wall has its self weight which is taken by plinth beam. It also provides uniformity to building at plinth level.

- **Procedure for constructing plinth beam -**

    1. Place where plinth beam would be laid is fixed and marked using the plinth beam plan given below.

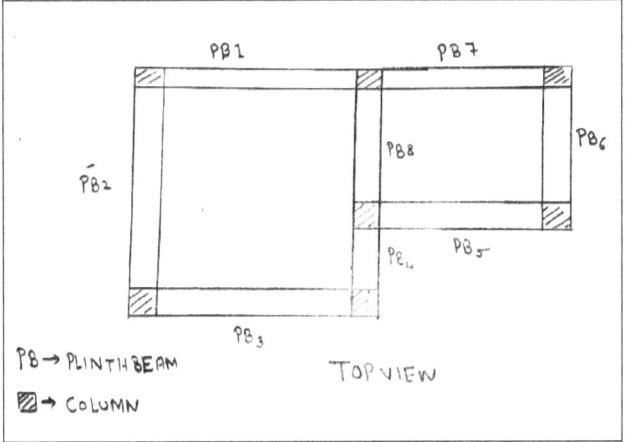

2. Cement sand paste is applied on the portion where the plinth beam is to be laid.

3. Now the steel frame work for the beam is to be done. (More details are given in slabs and beam chapter).

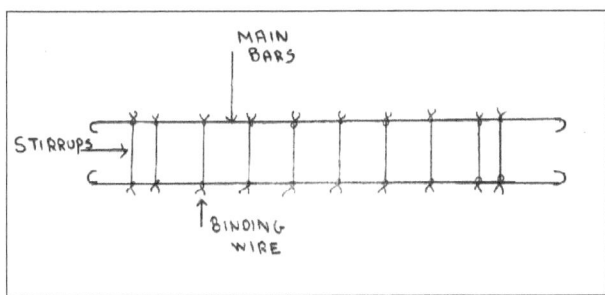

4. Wooden formworks are made of the needed dimensions and are placed at the fixed position.

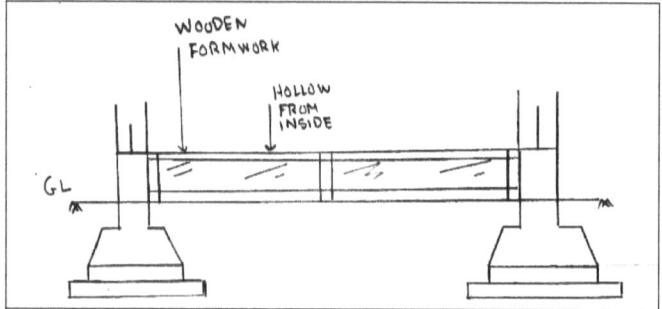

5. Steel frame work of the beam is placed in it as well, and the verticality and uniformity of the formwork is checked using plum-bob.

6. Concrete is poured into the formwork and is vibrated or tamped to remove the entrapped air and consolidate the concrete.

7. After 24 hours the wooden formwork is removed and the ready structure is watered.

8. The inner void spaces are filled with soil up to the height of upper surface level of plinth beam.

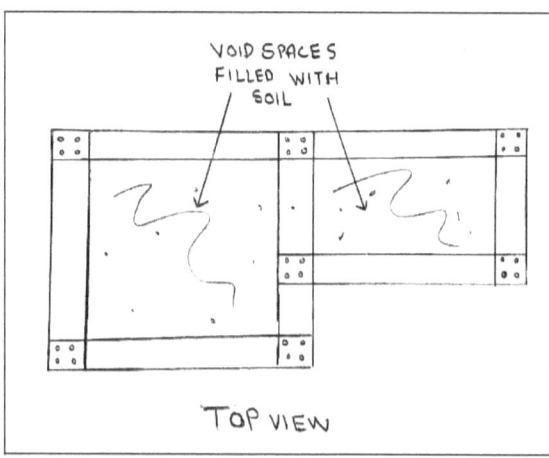

9. Ground slab is casted on which the flooring work in future is done.

If the house is needed to be raised from the ground level, then a brick wall is set up to the desired height and above this brick wall, plinth beams are constructed.

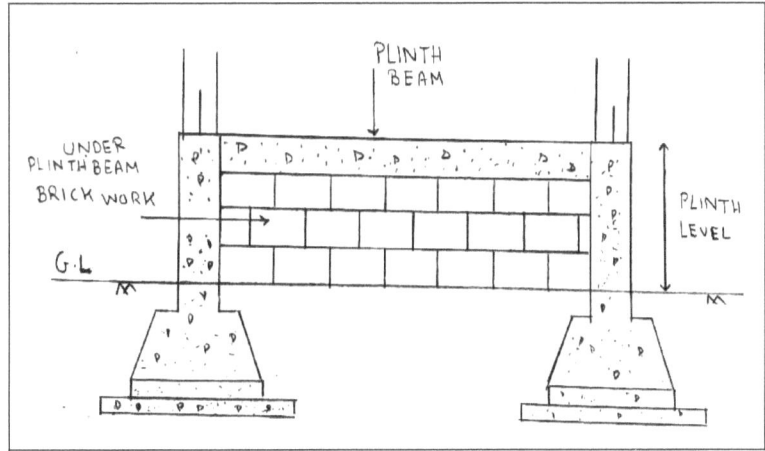

Chapter 7 – Column

As of now the construction up to floor on ground storey is done. Further we raise the house by erecting columns at stated spots.

Column is a vertical member which takes the complete load from the beam together with the self weight and carries it to the foundation. So basically column is a compression member as the burden proceeds along its longitudinal axis.

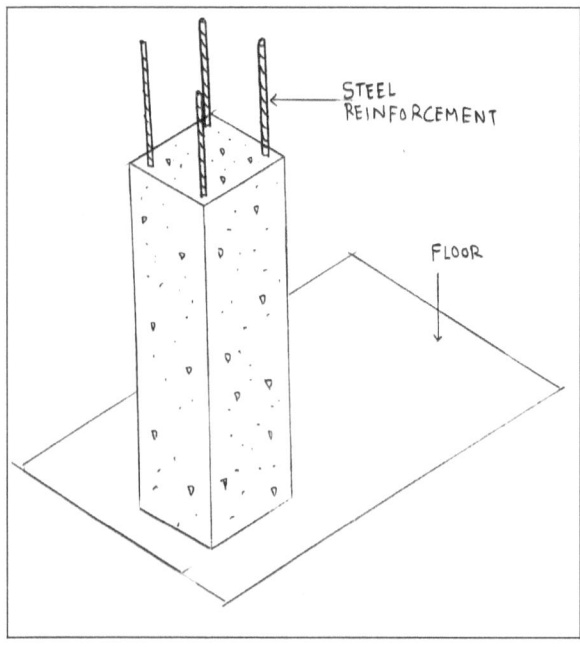

A whole superstructure stands with the assist of columns and building collapse comes about mostly due to column failure, hence it is also very crucial component.

❖ Classification of columns –

1. According to the shape –

The columns can be casted into non-identical shapes like square, rectangle, circular, hexagonal etc. The basic shapes are square and rectangular, as these two shapes are most general and serve only the load carrying role as every regular structural element.

The circular and hexagonal columns contribute to the aesthetics of the house i.e. to add quality view to the house. These shaped are put into services in the elevation of the house.

Generally square and rectangular columns are constructed in the inner parts of the house like bedrooms and kitchen. Whereas circular and hexagonal columns are used in open part of the house, like verandah.

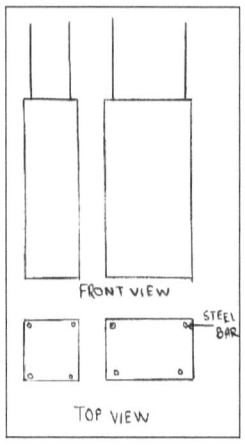
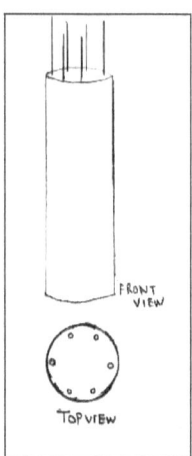
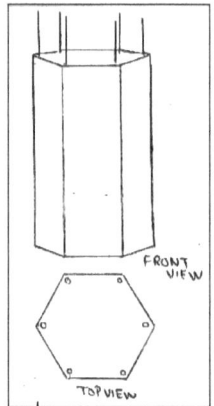

2. According to the length of column –

Based on the length, columns are classified as short column and long column.

- **Short column**

When the ratio of length of the column to the least lateral dimension (width) is less than or equal to 12, then it is called short column.

$$\text{Length}/\text{LLD} \leq 12$$

(LLD – least lateral dimension)

- **Long column**

When the ratio of length of the column to the least lateral dimension is greater than 12, then it is called as long column.

$$\text{Length}/\text{LLD} > 12$$

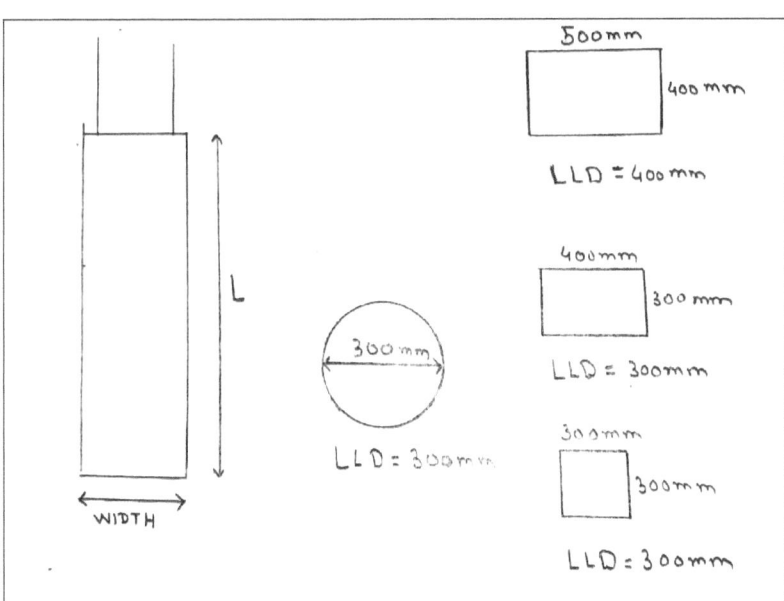

In circular columns the LLD is the diameter of the cross sectional area.

In normal 2 storey house the height of the column is mainly 3m (the floor to floor height). So the L/LLD ratio is always less than 12, hence the columns are always short. Long columns are used in buildings like malls, museum etc.

❖ Process of construction -

1. Column layout (Positioning) -

As the house construction nowadays is of column footing, so the position of columns are determined and fixed during the foundation work itself. The steel reinforcement sticking out from the foundation gives us the position of columns; we just need to extend the column from that steel work.

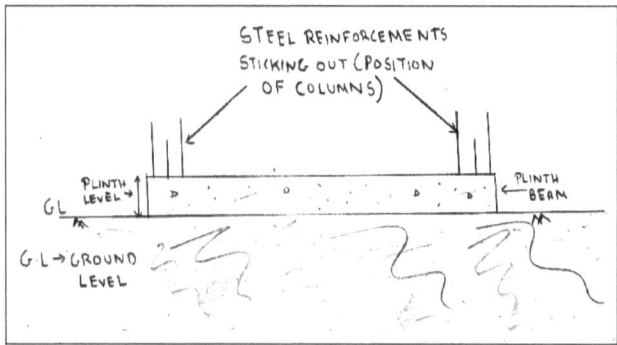

2. Reinforcement work (Steel framework) –

After determining the position of the column, it's time to prepare the steel cage of the specified shape and dimensions.

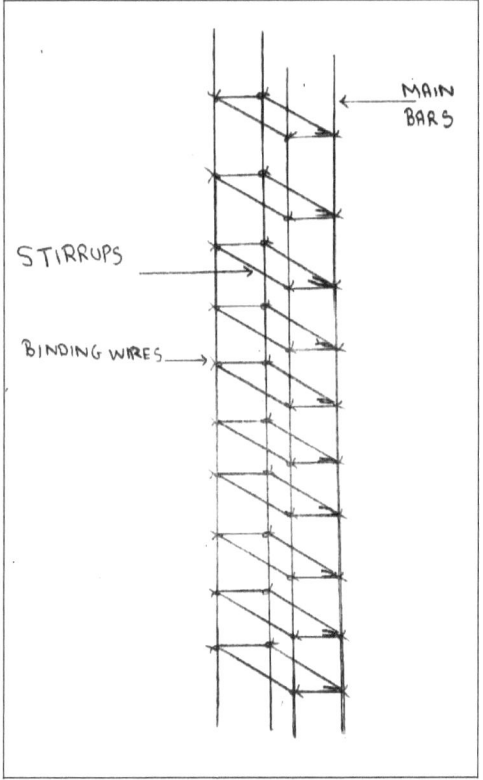

In regular square and rectangular columns, minimum 4 bars are necessary; while in circular and hexagonal, minimum 6 bars are needed to erect the column.

The horizontal distance between vertical rods in column should not exceed 300mm and should be greater than 5mm of coarse aggregate size. Clear cover for column is 40mm i.e. the steel bars should be 40mm inside the concrete surface. For this purpose cover blocks are positioned so that the steel gets dipped in concrete and do not advance towards the surface.

The vertical steel bars are held in vertical position by lateral ties called as stirrups. It is necessary to bend the stirrups at 45° angle after strapping the vertical bars; if not then the structure may lack in earthquake resistance.

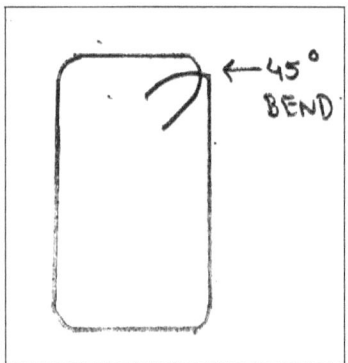

The number and diameter of steel bars, and the diameter and spacing between stirrups are listed in the design sheet.

For e.g.-

C1 – 6 – 20mm Ø and stirrup – 10mm Ø @130mm c/c

This means for column number1, 6 bars of 20mm diameter are required with 10mm diameter stirrups with 130mm spacing.

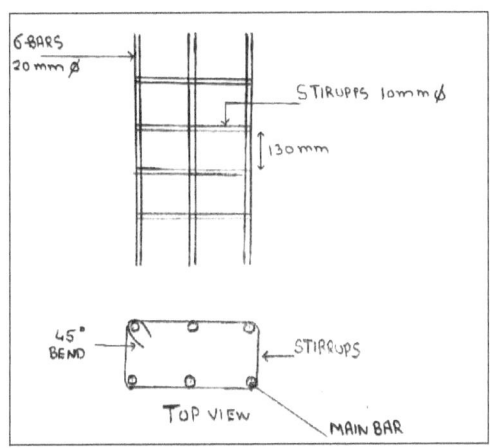

The whole steel network is binded using binding wires.

After the steel frame work is ready, the lower portion of the frame is lapped to the sticking out steel bars from the foundation, and the verticality is checked by using plum-bob (A lap is when two pieces of reinforcing bars are overlapped to create a continuous line or bar). The length of lap varies depending on the concrete strength, the steel bar grades, size and spacing. The lapping length is also listed in the design sheet of column.

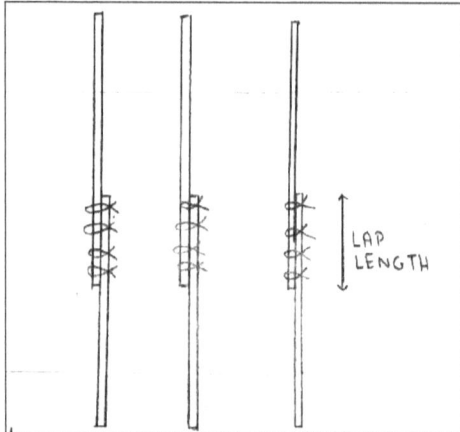

The stirrups needed are more in the lapping span hence the spacing between the stirrups are less in the lapping span to hold the frame properly.

3. Column shoe (starter) –

Column shoe also called as starter is a small fragment of column which is casted one day before the casting of main column. The cross sectional area of the starter is same as the column and the height is between 3 to 4 inches.

The only reason to build a starter is that it provides base support to the formwork which is needed to be fixed. If shoe is not provided then there is a chance of shifting of formwork while pouring and tamping wet concrete, resulting in dislocation of column structure.

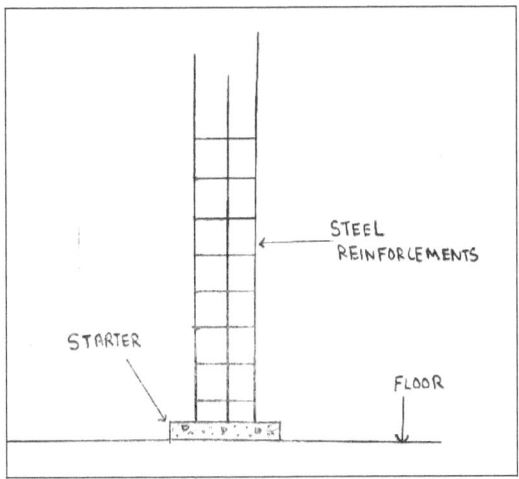

4. Formwork (shuttering) –

A formwork is nothing but a temporary mold made from planks. It is also termed as shuttering. It is generally made up of metal or wooden planks. These planks are arranged around steel reinforcement in such a way that it creates a hollow space into which viscous concrete is poured.

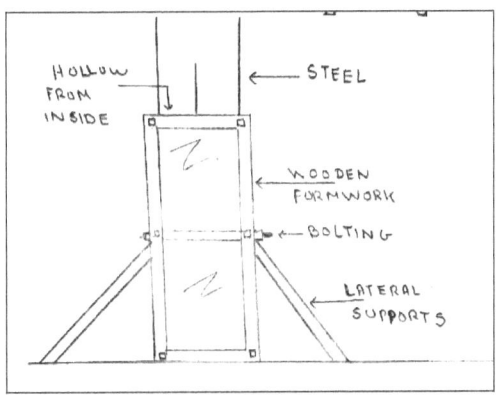

The shuttering should be anchored vertically by providing oblique supports. The diagonals of shuttering should be checked to ensure dimensional accuracy. Gaps at shuttering joints should be sealed and cover blocks placed.

5. Concrete work –

- **Preparing concrete –**

The concrete is prepared as per the design specifications mentioned. It is either manually mixed or by using machine. While mixing manually special care should be taken and the mixture prepared should be uniformly mingled.

A heap of cement, sand and aggregate is made and a depression in center is created using phawrah, in which water is poured and then mixing is carried out.

- **Pouring concrete –**

The concrete prepared is poured into the forms using pans. While pouring the height should not be greater than 1.5m as it may segregate the concrete. Proper vibration and tamping is needed so as to remove entrapped air and to maintain even surface. The column is casted up to the lower surface of slab.

Once the concrete is hardened, the formworks are removed. It is generally removed before 24 hours but not earlier than 16 hours of casting. Removing process should be done carefully without hammering; as it may damage the column surface.

6. Curing (Watering) –

Curing should be done at least for 15 days. In cold climate it should be done 3 to 4 times a day, and in hot temperature for 5 to 6 times a day.

The concrete structure is useless regardless to the grades of the concrete used, if curing is not done properly. For retaining water for longer period jute sacks are used.

Chapter 8 – Walls

Here is a freedom to choose; either we can go for brick masonry work or construct beams and slabs.

It is recommended to execute brick work first. Presuming we construct beams and slab first, then in this case there is a possibility of left out spaces between brick wall and beams at some places as per calculations. To dismiss this it is suggested to build a brick wall up to the calculated height and then cast beams and slab touching it.

❖ Size of brick –

The standard size of brick is 190mm×90mm×90mm i.e. 19cm×9cm×9cm.

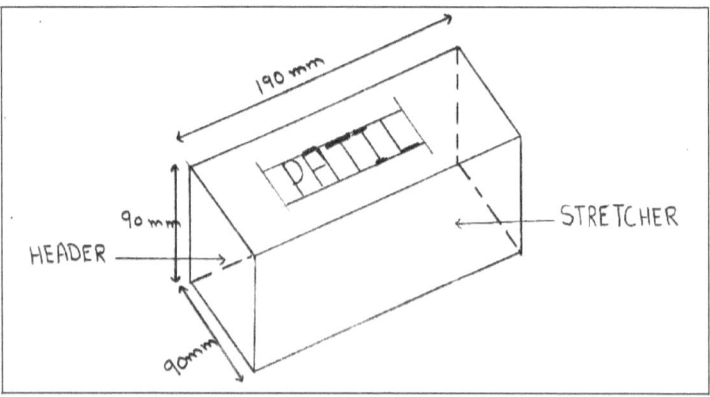

Purpose of frog – A frog is a depression on the top face of the brick provided to list the company name. The main motive of frog is to fetch and hold more mortar (binding paste) for better binding inter bricks.

❖ Characteristics of good bricks –

- ### Size and shape –

The brick should have uniform size and plan, rectangular surface with parallel sides and sharp edges.

- ### Colour –

The bricks should have a uniform deep red or cheery colour as indicative of uniformity in chemical composition and thoroughness in the burning of bricks.

- ### Texture and compactness –

The surface should not be too smooth to cause slipping of mortar. The brick should have pre-compact and uniform texture. A fractured surface should not show fissures, holes grits or lumps of lime.

- ### Hardness and soundness –

The bricks should be so hard that when scratched by a finger nail no impression is made. When two bricks are struck together, a metallic sound should be produced.

- **Water absorption –**

It should not exceed 20% of its dry weight when kept immersed in water for 24 hours.

- **Crushing strength –**

Should not be less than 10N/mm². I.e. 101 kg/cm².

❖ Types of bricks –

1. First class bricks –

These are the superior type of bricks which are perfectly burned, they holds all the characteristics of good bricks. They are regular in shape and size and have a uniform colour throughout. Also very durable and doesn't absorb more than 20% water of its own weight when immersed for 24 hours in clean water. The single brick of this type has a crushing strength of 10.29N/mm² i.e. 105kg/ cm².

2. Second class bricks –

These are slightly lower quality of bricks as compared to first class bricks. They are bit over burnt, and the shape and size also differs from standard form. The crushing strength of this brick type starts from 70kg/cm², and has a slightly greater water absorption value of 22% which is not allowable as per standards.

3. Third class bricks –

This type of bricks isn't uniform in size and shape. They can be over burnt or under-burnt. They don't absorb more than 25% water of its own weight when immersed 24 hours in fresh water. Crushing strength of this type is more than 30kg/cm².

❖

❖ Bonds in brick masonry –

Brick bonding is the ornamentation in which bricks are laid. It is nothing but the overlapping arrangement of bricks in order to tie the bricks together in a mass of brick work.

Depending upon the laying and bonding style of bricks, the 4 most generally used types of bonds are –

1. Stretcher bond.
2. Header bond.
3. English bond.
4. Flemish bond.

1. Stretcher bond –

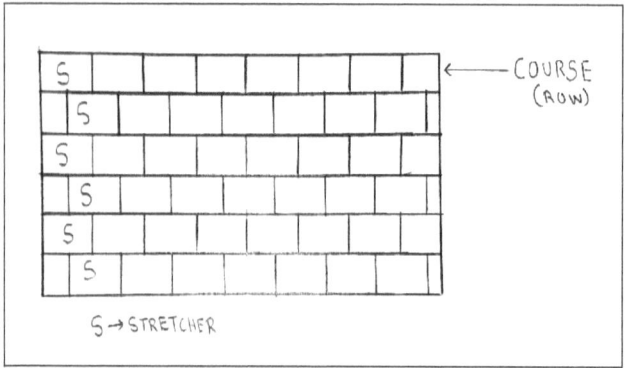

Stretcher bond, also known as running bond is the most commonly used bond in this age. It consists of all bricks laid as stretchers on every course with the course laid half bond to each other (2 courses overlaps at half bond). Stretcher bonds are easy to lay and causes less wastage.

2. Header bond –

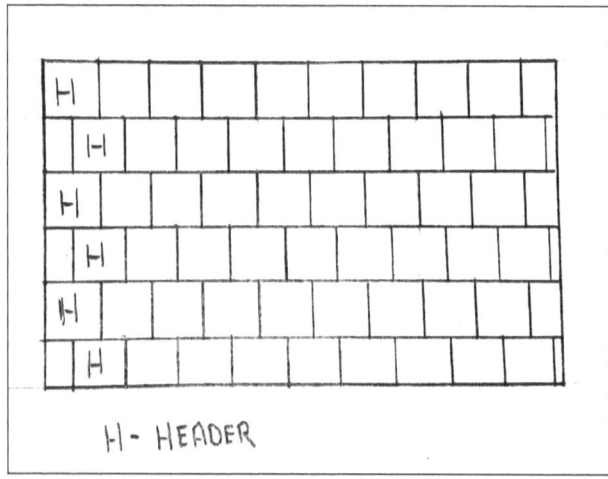

In this type of bonding, all bricks are placed as headers on the face of the walls. The overlapping is same as stretcher bond i.e. half bond.

3. English bond –

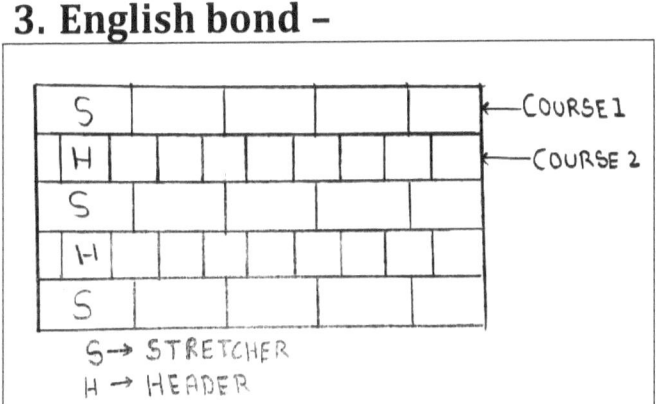

This type of bond has a course of stretcher only and a course of header parallel to it. I.e. it has alternate courses of headers and stretchers.

4. Flemish bond -

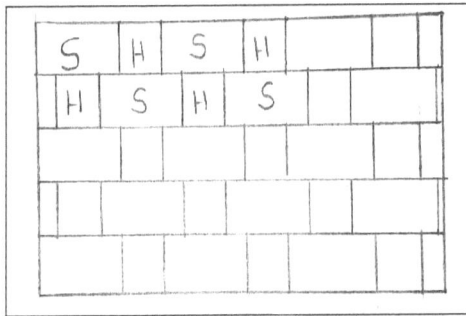

This type of bond consists of alternate headers and stretchers alternately in one course with the headers in one course being placed centrally over the structure in the course below. (I.e. the next course of the bricks is laid such that header lies in the middle of the stretcher in the course above).

Other types of bonds –

1. Dutch bond
2. Facing bond
3. English cross bond
4. Brick on edge bond
5. Raking bond
6. Zigzag bond
7. Garden wall bond

❖ Closers and bats –

- **Closer –**

Closer is a piece of brick cut longitudinally, and which is used to close up the bond at the end of the course.

Types of closers –

1. Queen closer

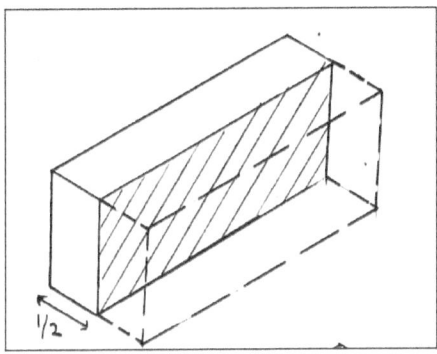

2. King closer

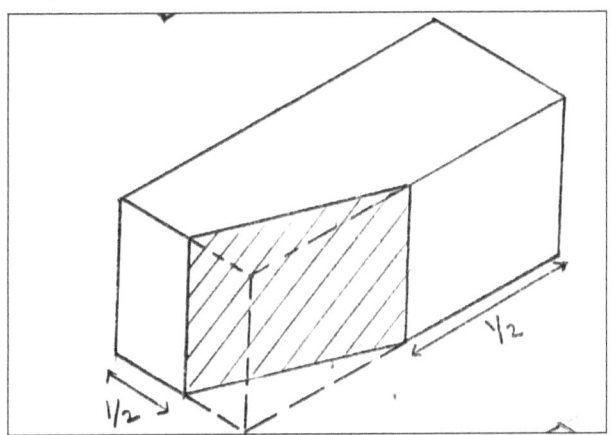

3. Beveled closer

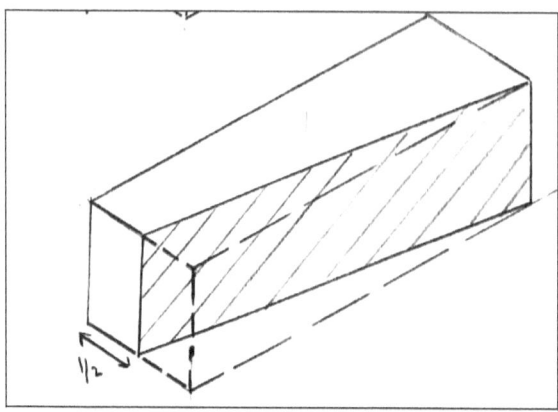

4. Mitred closer

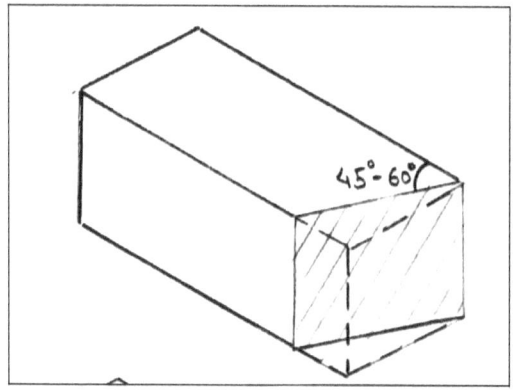

- **Bat –**

It is a piece of brick cut across the width.

Types –
1. Half bat

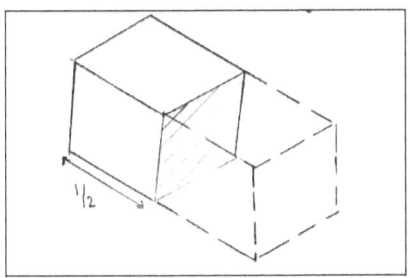

2. ¾ bat

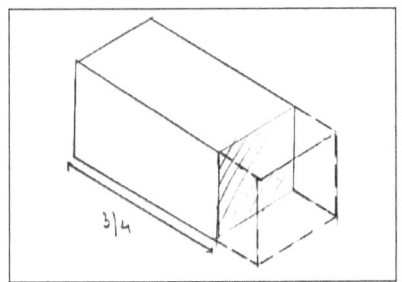

3. Beveled bat

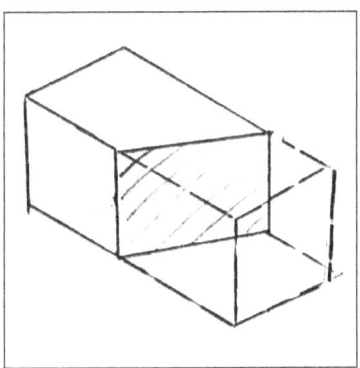

❖ Process of brick work –

- Before commencing masonry work, the wooden door frames are needed to be fixed, as brick walls are not laid there, and by doing this we secure the position of doors.

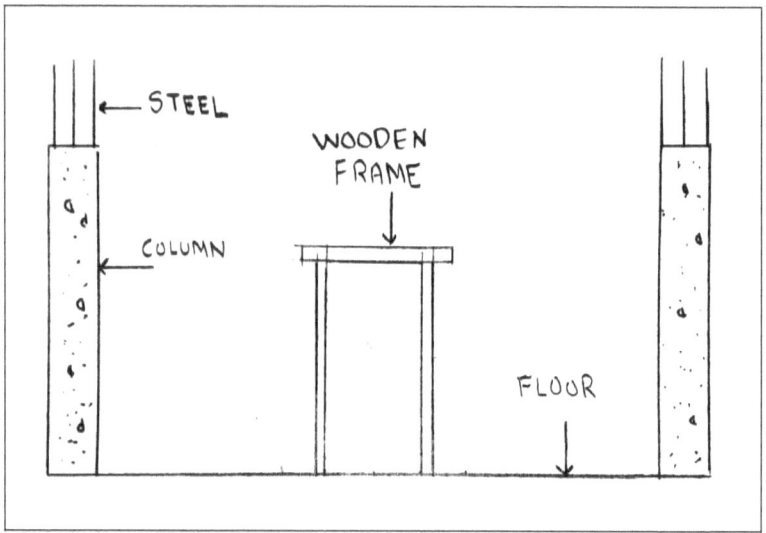

- The bricks are soaked in clean water for 4 hours before work to avoid absorption of water from the cement mortar (mixture of sand, cement and water i.e. sand cement paste).

- The line out is carried out for the entire area using steel tape.

- The 1st course of masonry shall be laid with great care, making sure that it is properly aligned, leveled and plumbed.

- The bricks for 1st course shall first be laid dry, (that is without mortar) along with a string tightly stretched between properly located cornerstones.

- The string shall then be stretched tightly along the faces of the two corner blocks and the faces of the intermediate ones adjusted to coincide with the line.

- The brick masonry shall be preferably laid in a composite mortar with mix ratio as mentioned in specification and drawing.

- It is either 1:4 or 1:6 (1 part of cement, 4 parts of sand

- Thereafter each brick on 1st course which has been laid dry shall be removed and re-laid over a bed of mortar.

- The thickness of mortar joints shall be 10 mm both horizontally and vertically.

- After every three or four blocks have been laid, their correct alignment, level, and verticality shall be carefully checked.

- The mortar shall be raked out from the joints with a trowel of each course and is laid to a depth of 10mm to 12 mm, so as to ensure the good bond for the plaster.

- When the height of the brick work reaches the sill level of window (the bottom level of window) i.e. 1m from floor, the temporary or permanent wooden window frames are needed to be fixed.

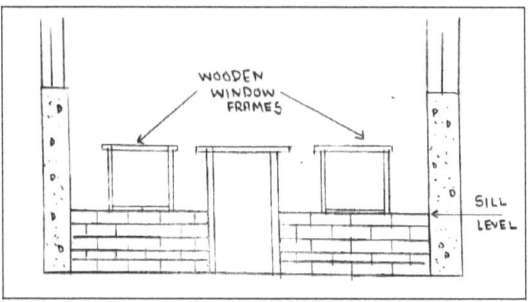

- Further when the brick work reaches the lintel level i.e. approximately 2m from floor level, the lintel RCC beams are casted above every doors, windows and openings.

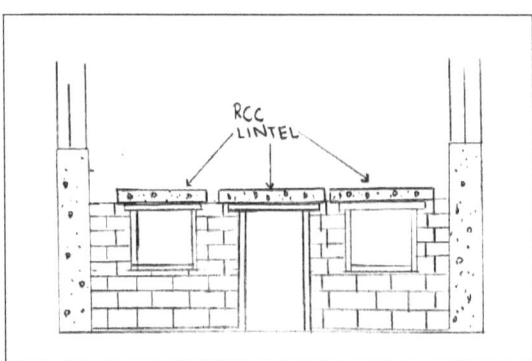

- Then the brick work is accomplished up to the desired height.

- The height of the wall to be done in a day's work shall be restricted to 1m.

- The external walls should have thickness of 0.23m, and internal walls of 0.15m.

- Scaffolding is needed for construction work at a height more than 1.5m.

- The brick should be cured (watered) for at least 2-3 weeks.

Lintel –

A lintel is a horizontal member which is placed across the openings like doors and windows. The basic purpose of lintel is to provide hard surface and support to the brick work above openings.

The width of the lintel beam is equal to the width of the wall, and the ends of the beam are anchored into the walls.

> **Scaffolding -**
>
> It is a temporary structure made of wooden planks, and bamboos (sometimes metal poles), used by workers while constructing or repairing something at a height beyond their reach.

❖ RCC Chajja

A chajja is the projecting or overhanging cover which is provided to protect from biased rain and unnecessary sun.

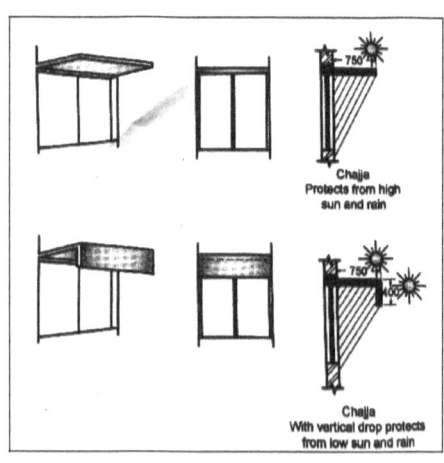

❖ Material calculation

Consider a brick work to be done of 10m³ with 1st class bricks and cement mortar of proportion (1:4)

The standard size of a brick

=190mm×90mm×90mm

=0.19m×0.09m×0.09m

The modular size of bricks

=200mm×100mm×100mm

=0.2m×0.1m×0.1m

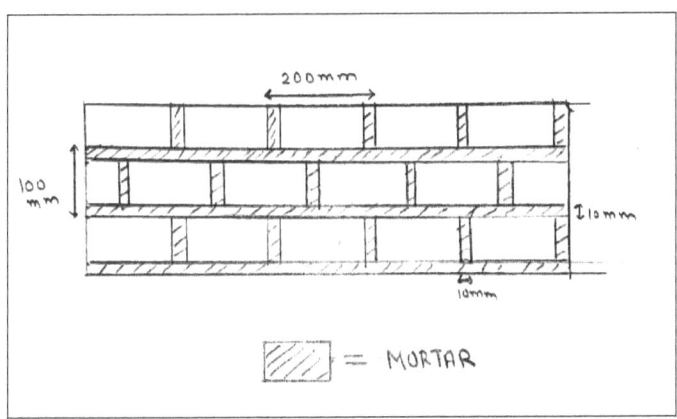

(Modular size of brick is the actual size of the plus mortar thickness)

No. of bricks

= volume of brickwork/ volume of modular brick

=10/0.2×0.1×0.1

=5000 nos.

Actual volume occupied by only bricks

= no. of bricks × standard size

=5000 × (0.19× 0.09× 0.09)

=7.695 m³

Volume of cement mortar

= Total volume − actual volume of only bricks

=10 − 7.695

=2.305 m³ (wet volume)

Dry volume increases by 33%

∴ Dry volume = 2.305 + (2.305 × 33%)

$$=3.065 \text{ m}^3$$

Considering 15% wastage

∴**Final volume of CM** = 3.065 + (3.065 × 15%)

$$= \mathbf{3.525 \text{ m}^3}$$

Materials required for cement mortar are – cement and sand (fine aggregate)

Ratio (1:4)

Quantity of cement = (Final volume / Addition of ratio)

× Part of cement

= (3.525/1+4) × 1

= **0.705 m³**

=0.705/0.035

$$= 20.4$$

$$\cong \mathbf{21 \ bags}$$

Quantity of sand = (Final volume / Addition of ratio)

$$\times \text{ Part of sand}$$

$$= (3.525/1+4) \times 4$$

$$= \mathbf{2.82 \ m^3}$$

Water cement ratio for masonry = 0.6

∴ Water/cement = 0.6

∴ Water/0.705 = 0.6

∴ Water = 0.423 m^3

∴ **Water = 423 litre .**

∴ For 10m^3 brick work it requires 5000units of brick, 21 bags of cement i.e. 0.705m^3 of cement, 2.82 m^3 of sand (fine aggregate) and 423 litre of water.

Currently, for masonry work concrete blocks are used. One advantage of blocks is that they form a stronger wall than bricks. A house built using blocks can have a last up to 100 years, while one built using bricks may last a bit shorter.

But the economy rate of brick wall is much less as compared to concrete blocks. A brick ranges from about rs-5 to rs-9 per piece and the starting price of concrete blocks is rs-40 per piece.

Concrete blocks are generally profitable to be use in huge non-residential buildings like commercial offices buildings, museums etc.

Chapter 9 – Staircase

Stairs are steps which are arranged in series so as to create circulation among different floors of a house.

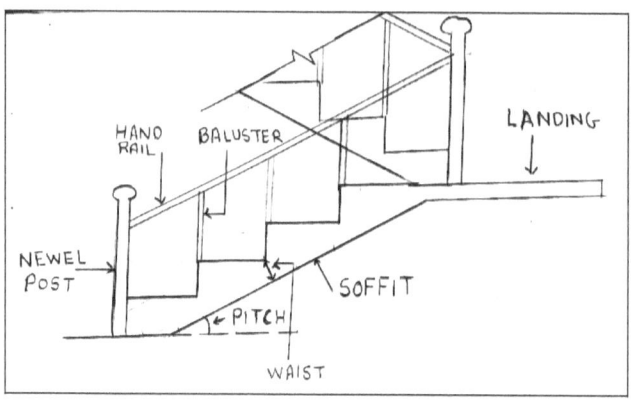

❖ Technical Terms

1. Tread –

It is the top surface of the step on which we can place our feet and climb.

2. Riser –

It is the vertical element which forms a space between two steps. Risers are responsible for the rise of steps.

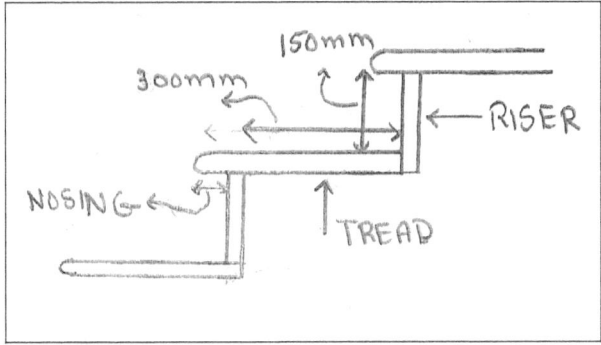

3. Landing –

It is an intermediate platform provided after a particular rise in steps. Landing is provided when stairs are needed to give turn in direction.

4. Flight –

Flight is a stairway between one floor and landing and the next.

5. Pitch –

It is the angle given to the stairway with respect to the horizontal.

6. Soffit –

It is the under surface of the stairway.

7. Waist –

The thickness of stairs RCC slab is called as waist.

8. Headroom –

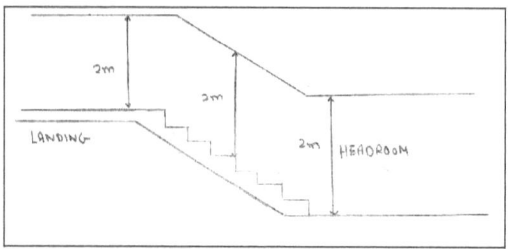

It is the room space available between the upper surface of tread and the ceiling above.

9. Hand rail –

Hand rail is an element which can ne gripped by hands so as to get assistance while climbing.

10. Newel post –

Newel post is the principle pole on which the hand rail is supported.

11. Baluster –

These are the secondary poles on which the hand rail is supported.

12. Nosing –

It is the small portion projected in tread to give good aesthetic appearance.

❖ Needs of residential staircase

• Location

The location of the staircase should be such that it should gain approximate access from every unit (rooms) of the house.

• Width of staircase

For dog legged staircase the width should be 1.8m (i.e. each flight should be of 0.9m) and 10cm i.e. total of 1.9m . 10cm is for the middle portion.

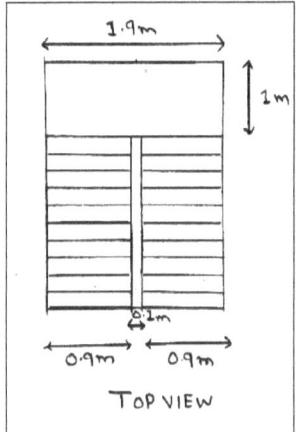
TOP VIEW

• Tread and Riser

Typically for residential building the rise of riser should be 150mm and tread width should be 300mm.

The floor to floor height = 3m = 3000mm

∴ **No. of risers** = 3000/150 = **20 nos.**

I.e. 10 risers on each flight.

No. of tread = No. of rise - 1

$$= 10 - 1$$

$$= 9 \text{ nos.}$$

No. of tread on each flight = 9 nos.

∴ **No. of tread** floor to floor = **18 nos.**

- **Headroom**

The headroom should be minimum 2m.

- **Landing**

The width of landing should be equal to the width of one flight or more than that is allowable, but not less. Generally it is taken as 1m.

- **Pitch of staircase**

The pitch should be between 42° and 25° for easy ascend and descend.

❖ Types of Staircase

1) Dog Legged

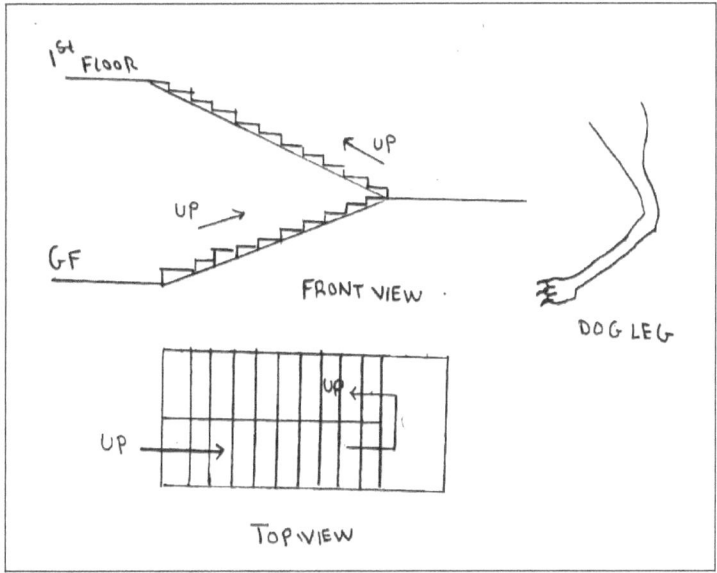

It is the most common type of stairs arranged with two adjacent flights running parallel. It has an abrupt change in direction of 180°. It is called as dog legged because when viewed across the section it appears like one.

2) Open well

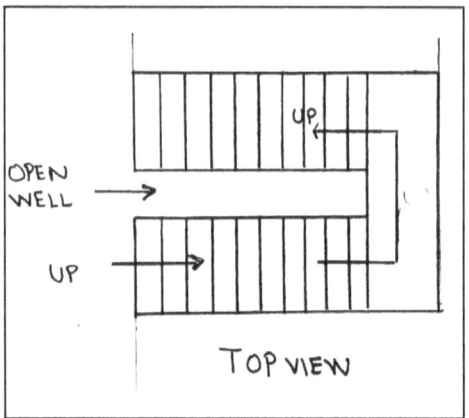

This type of stairs consists of an open space called as well, between two flights. This well is for good ventilation and natural lighting.

3) Quarter turn

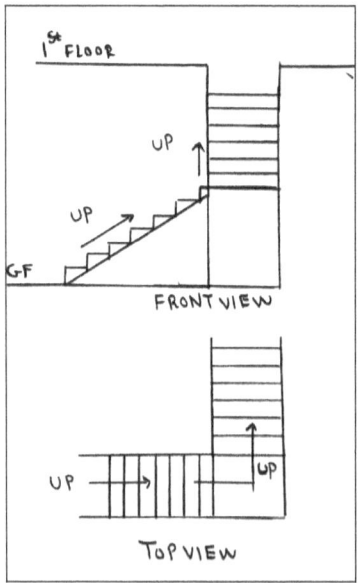

When the stairs take a 90° turn from the landing then it is called as quarter turn staircase.

4) Bifurcated

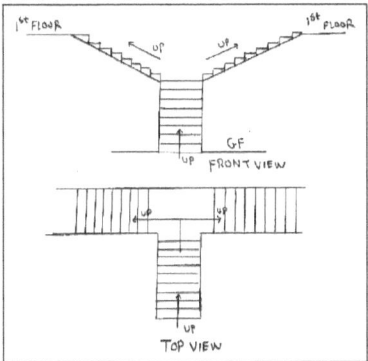

In this stairs, the flight is so arranged that there is a wide flight at the start which is sub-divided in two narrow flights at the landing. The narrow flights start from either sides of the mid landing.

5) Circular stairs

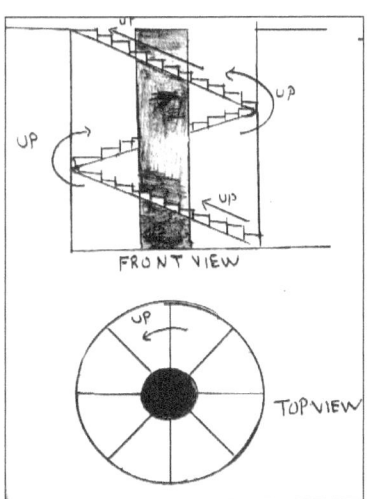

In this, all steps radiate from a newel or well hole, in form of winders. For RCC work the form works used should be flexible like softwood, cardboard etc.

6) 3 quarter turn

The direction of stairs changes 3 times with its upper flight across its bottom one.

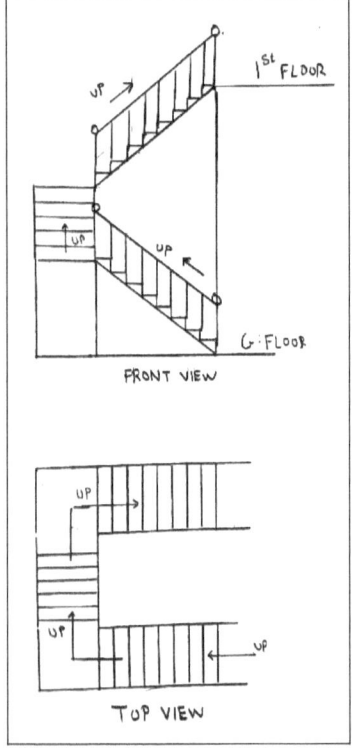

❖ Cross Section of RCC steps

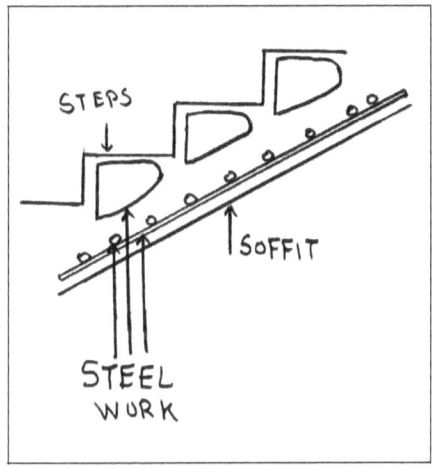

Chapter 10 – Slabs And Beams

Slabs and beams are two different components, but as they are casted together at one time, so are mentioned here together.

Sometimes staircases also are casted along with slabs and beams, but let's not make it jumbled and break the process down.

❖ Beams

Beam is the horizontal member of a structure, carrying transverse loads. Beams are generally rectangular or square in shape. The function of beams is to take load from slab and transfer it to the columns below.

❖ Types of beam

1) Simply supported beam

2) Fixed beam

3) Cantilever beam

4) Overhang beam

5) Continuous beam

1. Simply supported beam –

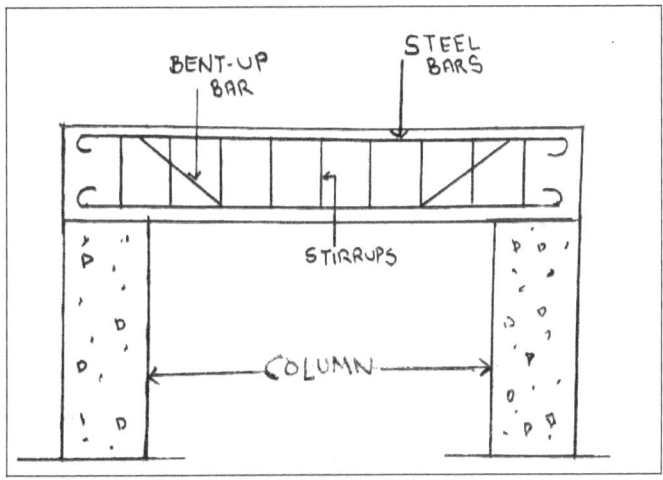

This type of beam is supported freely at the two ends on columns. In actual practice, no beam rests freely on the column.

2. Fixed beam -

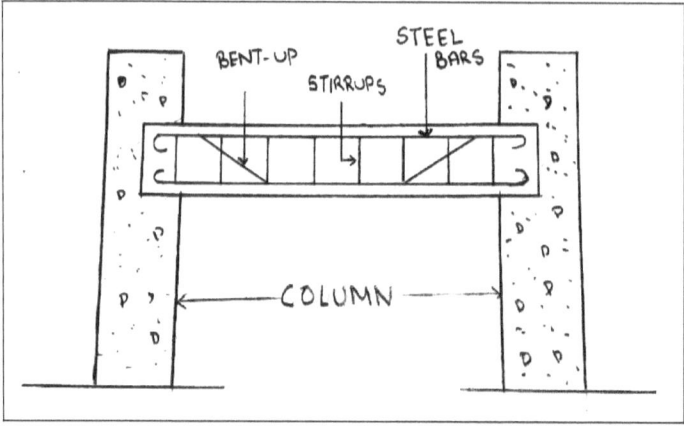

In this beam, both ends of the beam are fixed i.e. anchored into the columns including steel

3. Cantilever beam –

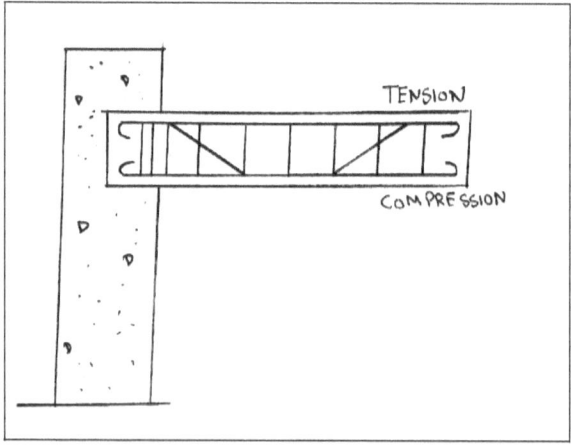

A beam is called cantilever when only its one end is fixed into the column and other is free.

4. Overhanging beam –

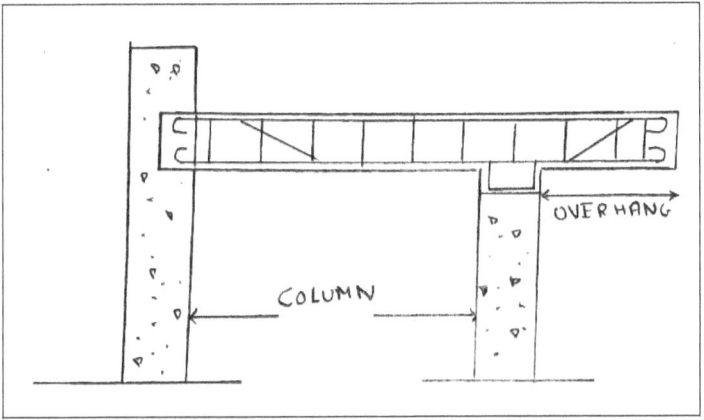

In overhanging beam, its end extends beyond the column support. Overhang is the unsupported portion of the beam.

5. Continuous beam –

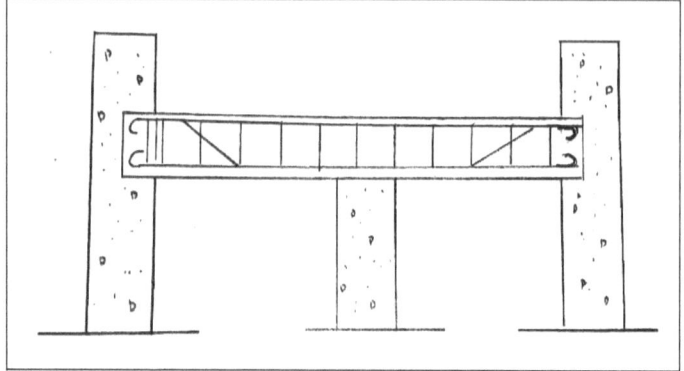

When the beam is supported on more than 2 supports i.e. columns, then it is called as continuous beam. They are used for longer span lengths.

❖ Slabs

Slabs are very common and important structural elements that are constructed to provide flat horizontal surfaces in building floors and roofs. The slab is supported on reinforced concrete beams which are casted monolithically with the slab.

❖ Types of slabs –

There are various types of slabs like flat slab, ribbed slab, continuous slab, simply supported slab etc. depending upon various factors.

But there are 3 important types of slabs depending upon the design criteria.

1. One way slab.

2. Two way slab.

3. Cantilever.

1. One way slab –

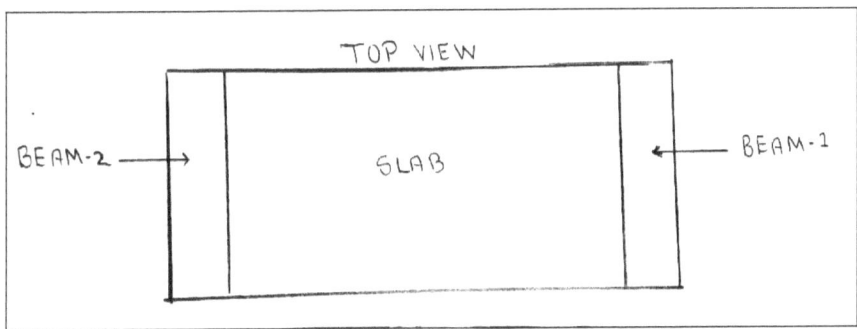

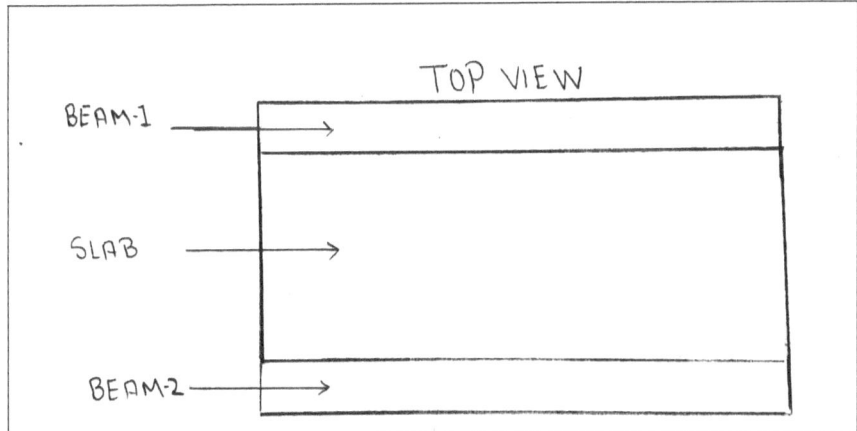

One way slab is a slab which is supported by beams on the two opposite sides to carry the load along one direction. The ratio of longer span (l) to shorter span (b) is equal or greater than 2, considered as one way slab because this slab will bend in one direction i.e. in the direction along its shorter span. However less size reinforcement known as distribution steel is provided along the longer span above the main reinforcement to distribute the load uniformly and to resist temperature and shrinkage stresses.

Longer span / shorter span ≥ 2

2. Two way slab –

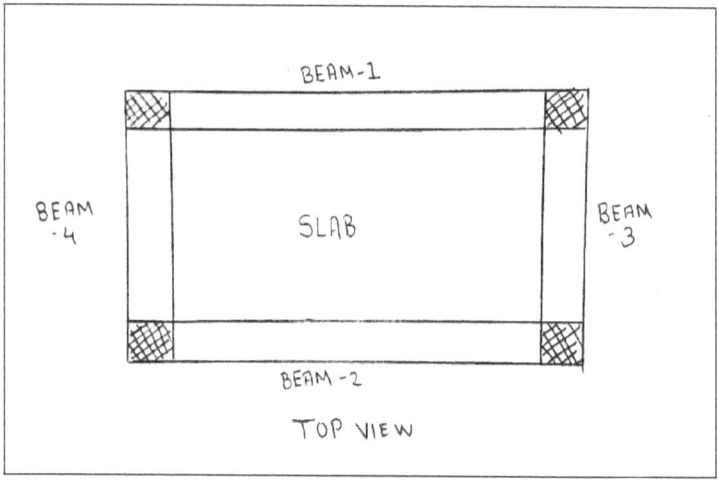

Two way slab is a slab supported by beams on all the four sides and the loads are carried by the supports along both directions, it is known as two way slab. In two way slab, the ratio of longer span (l) to shorter span (b) is less than 2. The slabs are likely to bend along the two spans, in this load is transferred in both the directions to the four supporting edges and hence main reinforcement is provided in both the directions.

Longer span / shorter span < 2

3. Cantilever slab –

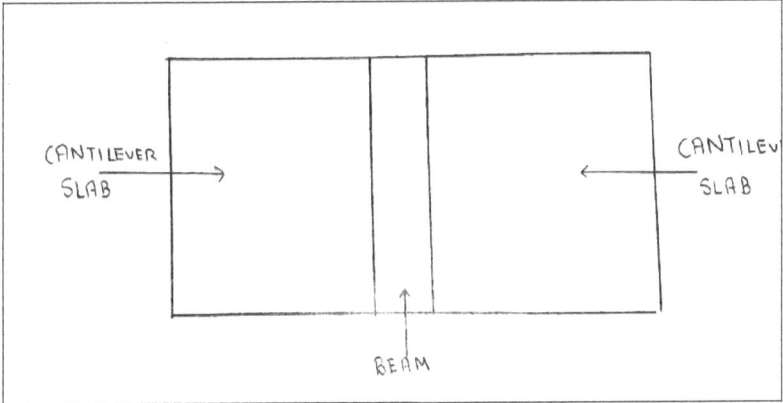

Cantilever Slab has only one support at one end and other three ends are open. See the image below "Cantilever slab". The main reinforcement of cantilever slab should be extended one and half times beyond its support.

❖ Process of construction

Laying of RCC beams and slabs can be done in 4 stages

1. Formwork
2. Steel reinforcement work
3. Pouring concrete
4. Curing

1. Formwork –

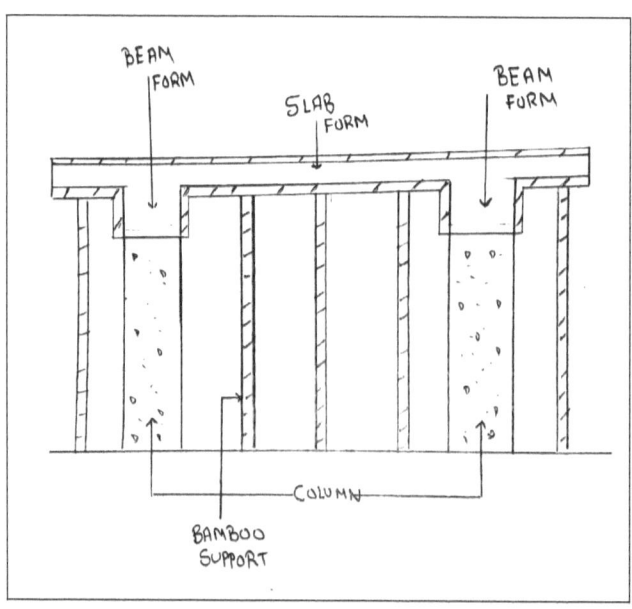

As mentioned earlier, formwork are temporary molds made of wooden or steel planks in which viscous concrete is poured to get settle and gain sufficient strength to be self supporting. The below fig shows the formwork of slab and beams.

2. Steel work –

Slab – At the time of designing the slab, it is considered that concrete is strong in compressive strength but weak in tensile strength. So to make the structure safe against tensile stress, steel bars are provided.

The steel work consists of preparing a steel grid by laying one set of bars in longitudinal direction and other in transverse direction tied using binding wires and elevated from the formworks surface using cover blocks.

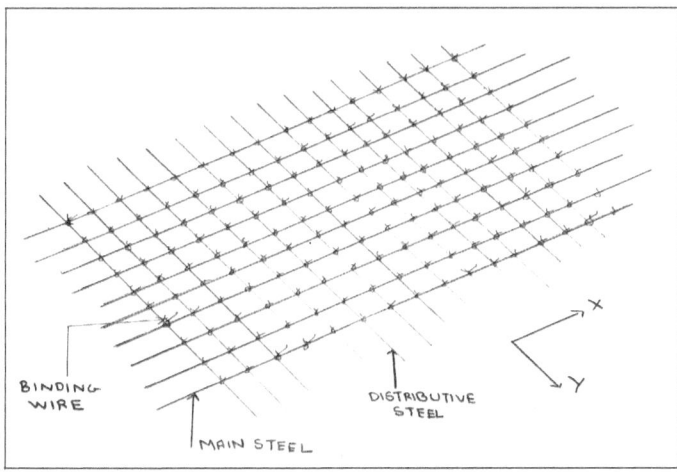

Beams - Beam's frame works are square or rectangular depending upon the design criteria which is been decided while designing. The framework of beams is similar to columns; the difference is that it is laid horizontally while in columns it is erected vertically. For beams to resist shear forces i.e. lateral forces, bent-up bars are provided, the bars are cranked at 45° from both sides.

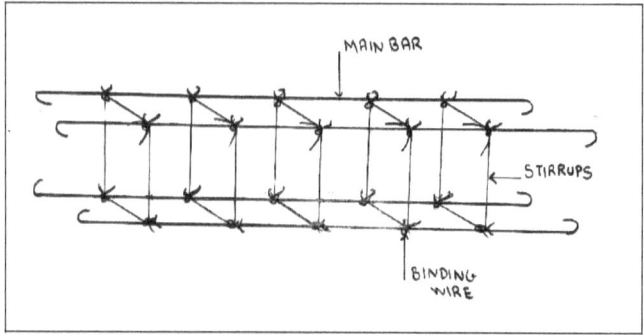

3. Pouring concrete

Concrete is prepared as per design specifications by mixing the materials properly using machine or hand mixing tools. It is then poured in to formworks with a falling height not more than 1.5m to avoid segregation. The poured concrete is then well compacted and vibrated using vibrators.

P.S. – Before pouring concrete, provisions for electricity points are made in the ceiling and beams.

4. Curing –

Curing should be done for 28 days for achieving full strength. For curing slabs ponding or spraying the surface with water is done. In ponding method, small rectangular or square artificial ponds are built with using bunds of clay or lean mortar or sand on the concrete surface. Water is filled in these small ponds 2 to 3 times per day depending upon the atmospheric conditions.

Chapter 11 – Plastering

Plastering is the covering with materials of various compositions applied either externally or internally to walls, ceiling etc. to cover rough walls and uneven surface of building.

Plastering is done by plastic mortar obtained by mixing cement (binding material) with sand (fine aggregate) and water in suitable proportion.

❖ Necessity of plastering –

1. To provide an uneven smooth, regular, clean and durable finished surface.

2. To resist the atmospheric influences particularly the infiltration of rain.

3. To cover inferior quality materials.

4. To prepare satisfactory base for decorating the surface.

5. To conceal the defective workmanship.

6. To fill the joints formed in masonry work.

7. The internal plaster provides a smooth surface which does not allow dust, dirt and vermin to lodge on it.

Cement plastering is commonly used as ideal coating for external and internal surface of wall. Cement plaster is usually applied in a single coat or double coat. Double coat plaster is applied where thickness of plaster is required to be more than 15 mm or when it is required to get a very fine finish. The process of applying a double coat cement plaster on wall surface consists of the following 4 steps.

1. Preparation of surface for plastering.
2. Ground work for plaster.
3. Applying first coat (or under coat or rendering coat).
4. Applying second coat (or finishing coat or fine coat).

Step 1 – Preparation of surface for plastering

1. Keep all the mortar joints of wall rough, so as to give a good bonding to hold plaster.

2. Clean all the joints and surfaces of the wall with a wire brush, there should be no oil or grease etc. left on wall surface.

3. If the surface is smooth or the wall to be plastered is old one, then rake out the mortar joint to a depth of at least 12 mm to give a better bonding to the plaster.

4. If the projection on the wall surface is more than 12 mm, then knock it off, so as to obtain a uniform surface of wall. This will reduce the consumption of plaster.

5. If there exist any cavities or holes on the surface, then fill it in advance with appropriate material.

6. Roughen the entire wall to be plastered.

7. Wash the mortar joints and entire wall to be plastered, and keep it wet for at least 6 hours before applying cement plaster.

Step 2 – Ground work for plaster

In order to get uniform thickness of plastering throughout the wall surface, first fix dots on the wall. A dot means patch of plaster of size 15 mm * 15 mm and having thickness of about 10 mm.

1. Dots are fixed on the wall first horizontally and then vertically at a distance of about 2 meters covering the entire wall surface.

2. Check the verticality of dots, one over the other, by means of plumb-bob.

3. After fixing dots, the vertical strips of plaster, known as screeds, are formed in between the dots. These screeds serve as the gauges for maintaining even thickness of plastering being applied.

Step 3 – Applying first coat or under coat or rendering coat

1. In case of brick masonry the thickness of first coat plaster is in general 12 mm and in case of concrete masonry this thickness varies from 9 to 15 mm.

2. The ratio of cement and sand for first coat plaster varies from 1:3 to 1:6.

3. Apply the first coat of plaster between the spaces formed by the screeds on the wall surface. This is done by means of trowel.

4. Level the surface by means of flat wooden floats and wooden straight edges.

5. After levelling, left the first coat to set but not to dry and then roughen it with a scratching tool to form a key to the second coat of plaster.

Step 4 – Applying second coat or finishing coat or fine coat

1. The thickness of second coat or finishing coat may vary between 2 to 3 mm.

2. The ratio of cement and sand for second coat plaster varies from 1:4 to 1:6.

3. Before applying the second coat, damp the first coat evenly.

4. Apply the finishing coat with wooden floats to a true even surface and using a steel trowel, give it a finishing touch.

5. As far as possible, the finishing coat should be applied starting from top towards bottom and completed in one operation to eliminate joining marks.

Step 5 - Curing

After completion of the plastering work, it is kept wet by sprinkling water for at least 7 days in order to develop strength and hardness.

❖ Requirements of an ideal plaster –

It should be smooth, non – absorbent, reasonably sound, deadening, flame retarding (which prevents or inhibits the outbreak of fire), washable and not affected by rise or fall in temperature. The plaster should not shrink while drying and setting. It should adhere firmly to the surface and should provide the surface with required decorative effect and durability.

❖ Material calculation –

Consider a plaster work to be done for 100 m² with Cement Mortar ratio (1:3) and a plaster thickness of 15mm i.e. 0.015m.

So total **wet volume** = Area of work to be

$$\times \text{ Thickness of plaster}$$

$$= 100 \times 0.015$$

$$= 1.5 \text{ m}^3$$

In order to convert dry volume from wet volume, quantity should increase by 33%.

∴ **Dry volume** = 1.5 + (1.5 × 33%)

$$= 1.995 \text{ m}^3.$$

Consider wastage of materials as 15%

∴ **Final volume** = 1.995 + (1.995 × 15%)

$$= 2.294 \text{ m}^3.$$

Material required for plastering work are cement, sand and water.

Quantity of cement = (Final volume / Addition of ratio)

× Part of cement

$$= (2.294 / 1+3) \times 1 = \mathbf{0.57 \text{ m}^3}.$$

$$= 0.57 / 0.035$$

$= 16.25$ bags

$\cong$ **17 bags**

Quantity of Sand $=$ (Final volume / Addition of ratio)

$\times$ Part of sand

$= (2.294/ 1+3) \times 3$

$= 1.72 \text{ m}^3.$

Water cement ratio for plastering $= 0.6$

$\therefore$ Water/Cement $= 0.6$

$\therefore$ Water/$0.57 = 0.6$

$\therefore$ Water $= 0.342 \text{ m}^3.$

$\therefore$ **Water $= 324$ litre.**

So for 100 m² plastering work it requires 17 bags of cement i.e. 0.57 m³ of cement, 1.72 m³ of sand (fine aggregate) and 324 litre water.

Chapter 12 – Plumbing

Plumbing works on the simple concept of "water in -- water out." In a new home, the plumbing system features three main components, the water supply system, the drainage system and the appliance/fixture set.

Local codes determine standard plumbing procedures, but a new home's fixture placement, pipe routing diagram and pipe size depends on the home's individual layout.

❖ Water supply system(Water- in)-

The purpose of the supply system is to convey clean usable water from the water source to the different parts of house like bathroom, kitchen, toilet etc.

The first requirement of supply system is to have water storage facility. This is achieved by 2 storage tanks, one which is underground storage tank and one overhead storage tank.

These water storage tanks are necessary because it is not feasible to connect the municipal water main directly to the small units in the house like taps, showers etc.

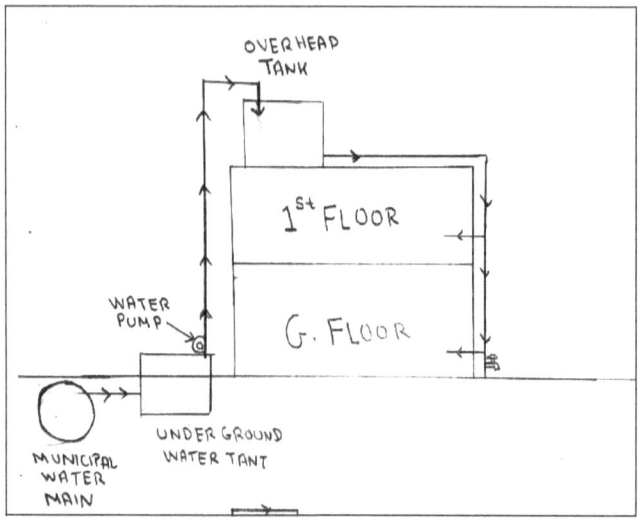

The overhead tanks are mounted on terrace above staircases. Whereas, underground storage tank should be situated in the rear side of the house.

The municipal water which is supplied is stored in the underground storage tank, from which the water is further pumped to the overhead tank, and the supply system terminates by conveying the water from the overhead tank to the different house units.

---Start To End On How To Build A House---

Municipal water main

⇓

Underground storage tank

⇓

Overhead storage tank

⇓

House units (taps, showers etc)

- **Ground water**

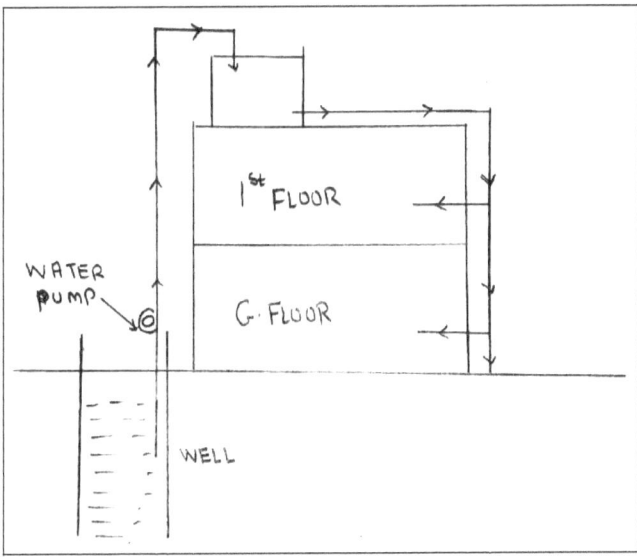

In case of availability of clean refined source of ground water, there is no need for the supply of municipal water. The water is made available by digging wells or bore wells, which is pumped to the overhead tank and further conveyed to the different units of the house.

- ## Types of pipe used in plumbing

1. Cast iron pipes (CI)
They are cheaper and highly durable.

2. Galvanized iron pipes (GI)
Mostly used for water supply work inside the building.

3. Steel pipes
They are used when the water pressure is more than 7kg/sq cm. they can have large diameter.

4. PVC (poly vinyl chloride) / Plastic/ Polythene pipes
They are light in weight, non corrosive, low in cost, does not require threading.

❖ Drainage System (Water-out) -

Now, when the water is conveyed to the different house units, it is used, and the considerable waste water is needed to be drained off. For this purpose, drainage system is needed.

The drainage system collects the waste water from kitchen, sink, toilet and bathrooms etc; and withdraws it to the sewer lines via septic tanks.

❖ Types of drainage system -

There are 4 types of pipe system for draining work, namely

1. One pipe system
2. Two pipe system
3. Single stack system
4. Single stack partially ventilated system

1. One pipe system

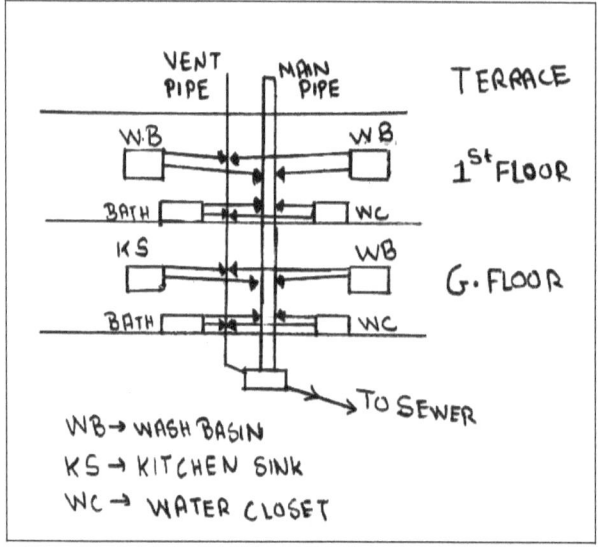

In this system a single pipe is used to carry the waste water from all the units, and a ventilation pipe is provided to all the traps of WC, bath, sinks and also to the gully trap.

2. Two pipe system

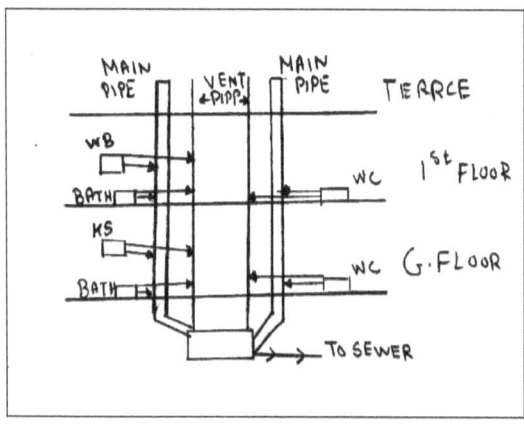

In this, two separate pipes are provided, one to carry solid waste water from WC and other to carry waste from bath and sinks. A separate vent pipe is imparted for each main pipe, hence making a total of 4 pipes in altogether.

3. Single stack system

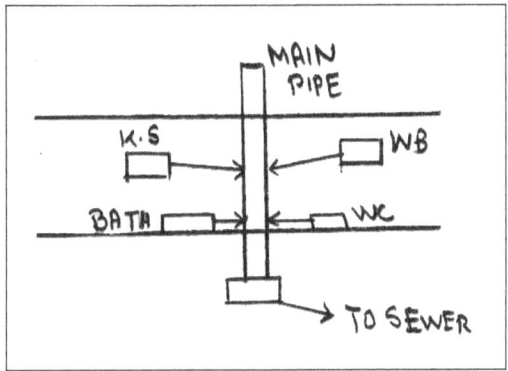

Here, only a single pipe is provided which also acts as the ventilation pipe.

4. Single stack partially ventilated system

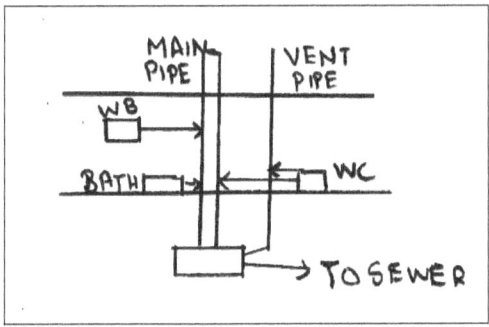

It is similar to one pipe system but as the name suggests, it is partially ventilated i.e. the ventilation pipe is provided only to WC traps and not others.

The water from pipes advances into the gully trap, which is then taken to the septic tank and finally discharges to the sewer lines.

For making the discharge process of waste water trouble-free, the pipes conveying waste water from septic to the sewer lines, are laid at a slope and not horizontally. By this the gravity helps in discharging the heavy materials and avoids clogging and blocking of pipes.

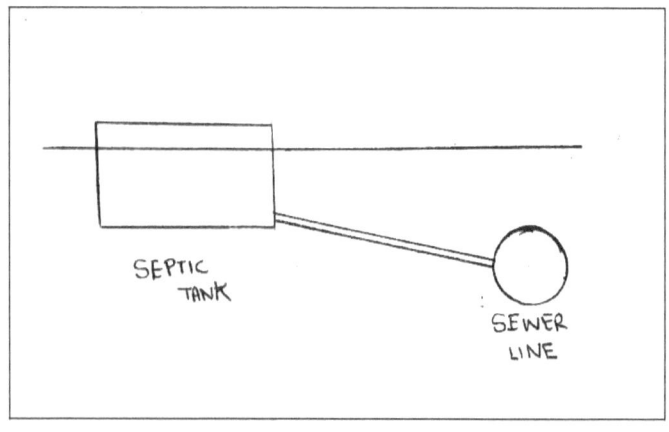

❖ Sanitary units

1. Wash Basin –

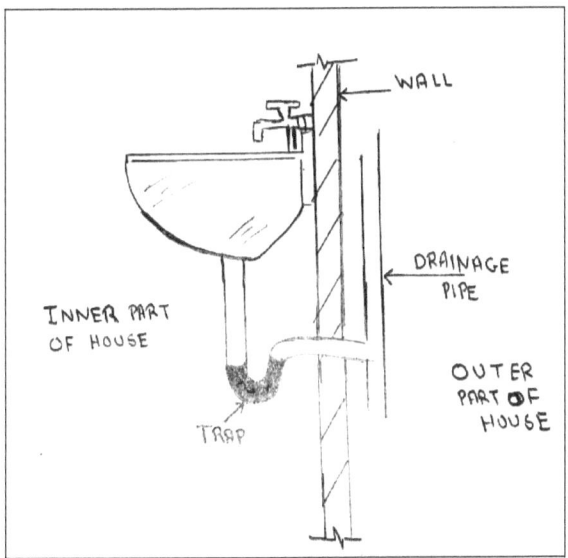

Wash basins are made up of white glazed earthenware. They are available in various shapes like square, rectangle, circle etc. the dig of wash basin is given below.

2. Bathroom –

- **Bathtub** – Bathtubs are mostly pre-casted and made up cement concrete with marble finishing.

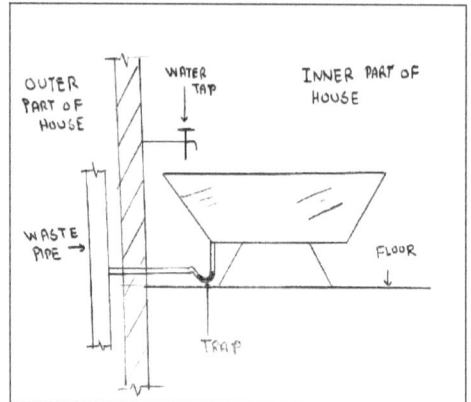

- **Normal bathroom** - This consists of a drainage hole which is provided to the approximation of the waste pipe.

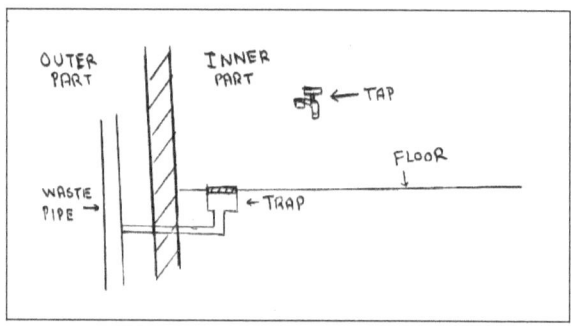

3. Water closet (Toilet) –

- **European style** – This is made up of porcelain. It is made for providing a seat while defecation. It is above the floor to a height of 0.5 – 0.6m.

Indian style – this is also termed as squat toilet. It consists of a bowl which is fixed into the floor and has a depth of about 1 foot to 1.4 feet. So the floor of WC here should be constructed at a height of 1 foot to 1.4 feet from the house floor level. Or should have the required depression while casting slab.

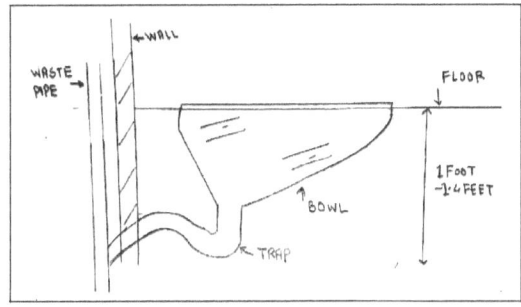

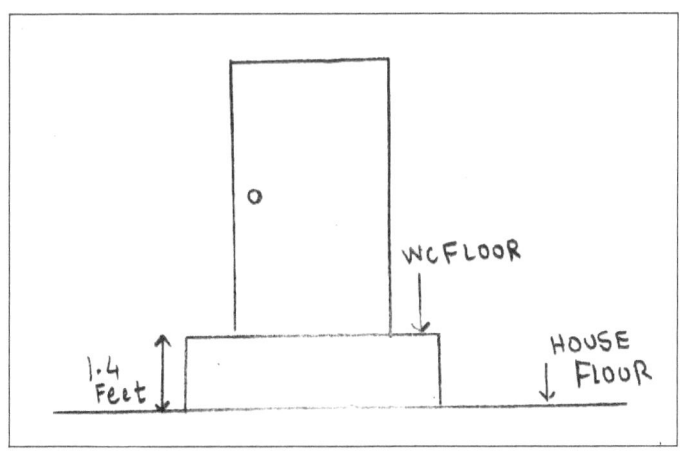

❖ Traps

A trap is a utensil which has a shape that uses bending path to capture water to prevent sewer's foul gases or foul air from entering the house and only allowing waste to pass through it.

- **Types of traps –**

1. Post on shapes
a) P – trap

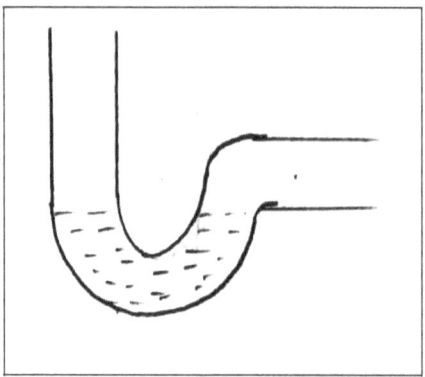

b) Q – trap

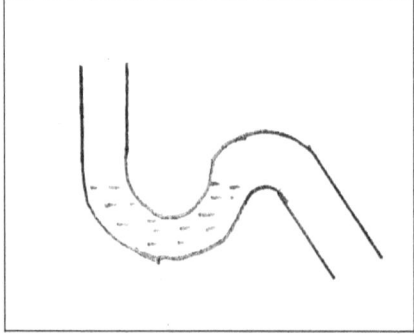

c) S – trap

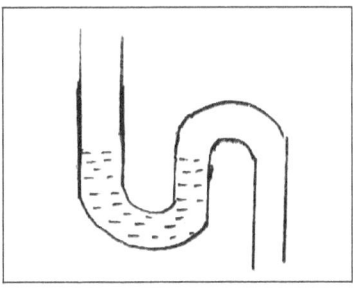

2. Post on use

a) Floor Trap (Nahni trap)

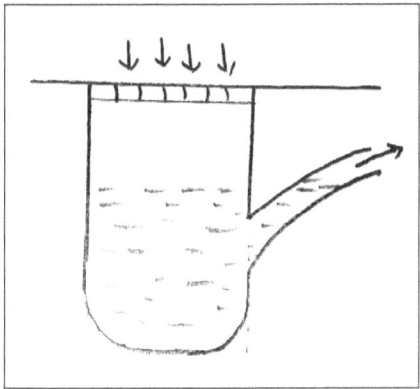

Floor trap also called as nahni trap is provided for all places in bathroom, sinks, kitchen, etc. it is generally made up of metal.

b) Gully Trap

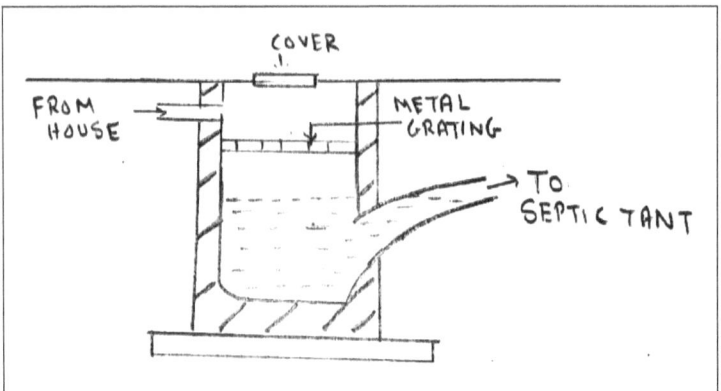

This trap prevents ingress of foul air, insects and vermin from sewers into house and resist spread of disease. It is provided outside the building. It is made up of masonry work with a metal grating on top.

Chapter 13 – Wiring

The next step after plumbing is to electrify the house for providing essential power required by the various electrical appliances like fans, lamps, television, air conditioners, fridge and others.

An electrical wiring laying out is done to the exact points at outlet of fans, lamps and other appliances as per the requirement in the house, out of which fans and lamps are mandatory. In every room there must be at least one fan.

1. Living room –

Living room is needed to have 3 sockets of 5A and a 15A socket; and should have illumination of 150-200 lux.

2. Kitchen –

Kitchen should have a 5A socket and 2 – 15A sockets for fridge and mixture grinder; and should be illuminated up to 200 lux.

3. Bedroom –

In each bedroom there should be 3 – 5A sockets and a 15A socket for air conditioner, with a 100-200 lux illumination.

4. Bathroom –

A bathroom requires illumination of 80-100 lux and demanded a 5A socket and a 15A socket for washing machine, inside or near it.

5. Study room-

It should comprise of 2-5A sockets and with a high illumination of 250-300 lux.

6. Verandah –

It is required to provide 5A sockets per every $10m^2$ of area and a 15A socket overall. The lights illumination should be adequate enough to light the verandah in night time.

lux – lux is a unit of illumination. It is a measure of the amount of light provided by something.

5A – 5 amp sockets are small round pin socket for primary appliances.

15A – 15 amp sockets are large round pins sockets similar to 5A.

Chapter 14 – Painting

We arrive on the scene once the plumbing and electrical have been run through the house and the drywall work has been completed.

At this point no other installations have been put into the house, such as flooring, kitchen and bathroom cabinets, toilets, furniture, home appliances etc.

❖ Process of painting –

There are no strict rules to follow the below process, but this is what the best I found.

Painting is done in four steps-

Surface preparation

Applying primer

Applying putty

Applying paint

Step1 - Surface preparation

Surface should be minimum 45 days old, well cured and completely dry. If there is any defect repair that with cement-sand mortar that is called patch work. Cure and dry the patch work. Apply a coat of lime-wash then allow it to completely dry. Then remove loose mortar, dirt or any other foreign material by applying sand paper on the surface.

Step2 - Applying primer

The main job of primer is to provide adhesion between the surface and the paint film. It also makes the surface smooth, less absorbent and increases the spreading ability of paint. Normally primer is applied by brush or roller. Before applying primer it is thinned with water.

Step3 - Applying putty

Putty is applied to repair cracks and to make smooth & level the surface. After applying putty the surface should be allowed to dry for 4 days. And then scrap off the surplus putty with sand paper. Preparation of putty: Putty is prepared by mixing 4 litre plastic paint, 1 litre enamel paint and 25 kg chalk powder with water.

Step4 - Applying Paint

Paint is applied two or three coat on the surface. After completely drying up the putty, the first coat of paint is applied by roller. The paint is thinned by mixing water. The water should be maximum 20% for 1st coat and maximum 15% for subsequent coats. After applying 1st coat surface should be allowed to dry for minimum 7 days before applying 2nd coat. And then 2nd coat of paint is applied on the surface. In this stage, if the surface is not smooth, luster, good opacity and spotless then the 3rd coat should be applied.

Once we have finished all our painting we do a final inspection of the home to make sure everything is complete and our job is done.

Chapter 15 – Flooring

Flooring is the general term for a permanent covering of a floor, or for the work of installing such a floor covering. Floor covering is a term to generically describe any finish material applied over a floor structure to provide a walking surface.

Flooring is mostly done by tiles which give good and pleasant finish. Tile is a nice addition to any basement. It protects against moisture and provides a visual contrast to the rest of the house.

❖ Types of flooring

1. Ceramic tiles
2. Marble
3. Mosaic tiles
4. Kotah
5. Granite
6. Shahabad stone
7. Quartzite tiles

8. Wood Look tiles

9. Porcelain tiles

----

1. Procedure of laying tiles
Step1 - Cleaning the floor

The floor on which the tiles are going to be laid should be cleaned properly.

Step2 – Finding the center point of the room

The center point of a particular room is found out by intersecting two straight lines initiating from the center of each side of the room. This is generally done by stretching chalk coated strings and then snapping it on the floor.

Step3 – Placing tiles without mortar

Before starting the project, start with a dry run, laying out the tiles and the spacers. This determines where to start the process and helps determine the width of the grout lines.

Step4 – Snap another chalk line

Snap an additional chalk line the width of a tile out from each wall. This will help keep the tile placement perfectly straight. Continue laying out tiles and spacers until placement is correct.

Step5 – Prepare mortar

Use a self-mix thin-set mortar for the tile installation. Pour just about an entire bag of the dry mixture into a large bucket. Add just enough water to get the dry mixture wet, and begin mixing. Continue to mix until the texture is creamy. Then let the mortar stand for about 10 minutes to get tacky.

Step6 – Apply the mortar

When the mortar is ready, begin working on one section of the floor. Spread the mixture on a 2'x2' section of the floor and use a notched trowel to obtain an even layer of mortar. Work in small sections to keep the mortar from drying before the tile is in position.

Step7 – Placing tiles

Place the tiles on the position with care and hammer gently with your fist to paste it.

Step8 – Cutting excessive portion

When you get to a wall where a standard tile will not fit, mark and make cuts with a standard tile cutter. If you don't use a tile cutter, mark the tiles and have a tile supplier cut them prior to installation.

Step9 – Placing spacers

Continue the process, using the chalk line as a guide and placing spacers between each tile to ensure uniform distance between the tiles. When all the tile work is complete, allow the tiles to dry in place for several days before grouting.

Step10 – Apply the grout

Grout is available in a variety of textures and colors. Pick a color that matches the color of the tile. Use a rubber trowel or float to spread the grout across the tiles at an angle to be certain to get it between each tile.

Grout is a particularly fluid form of concrete used to fill gaps. Grout is generally a mixture of water, cement, and sand, and is employed in pressure grouting, embedding rebar in masonry walls, connecting sections of pre-cast concrete, filling voids, and sealing joints such as those between tiles.

Step11 – Remove Excess grout

When the gout is in place, wipe away the excess with a damp sponge or cloth. Repeat this process several times, being careful not to remove the grout lines around the tiles.

❖ Skirting

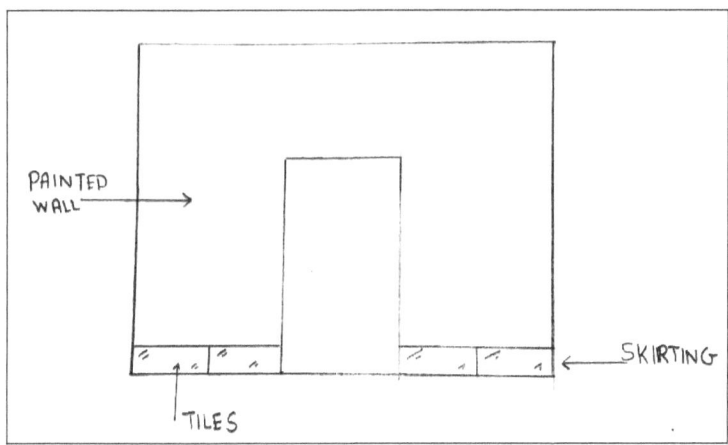

Skirting are the tiles that are laid in the corner of a wall and floor junction. It is mainly used to protect the painted wall bottoms while cleaning the floor with mops without leaving the mark on the wall bottom, thus it protects the wall from moisture. Skirting is provided for a height of about Six to Eight inches.

❖ Dado

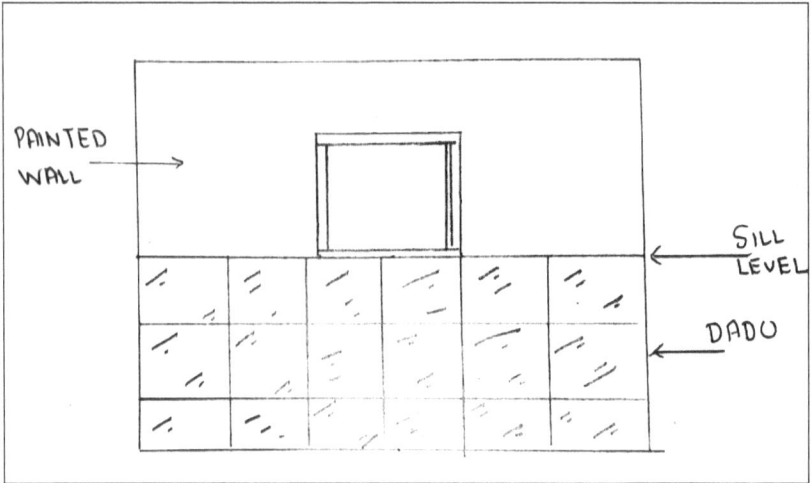

Dados are tiles applied from the floor level and for a height up to sill level of window in kitchens, toilets and bathrooms. They are applied for protecting the walls from splashing water in bathrooms and toilets, and to avoid stains on kitchen walls.

Chapter 16 – Doors and Windows

❖ Doors

These are open-able barriers hinged from one side in the wall opening. The function of a door is to give access to house and to different parts of the house, and to deny access whenever necessary.

The size of the door should not be less than 0.9m × 2m. Large doors may be provided at main entrance while minimum sized doors as 0.75m × 1.9m are used for bathroom and water closets.

- **Main Door**

The main door should add on some good architectural effects to the house. It can be single shutter or double shutter, and the type generally used for main doors is panel door.

---Start To End On How To Build A House---

----Start To End On How To Build A House----

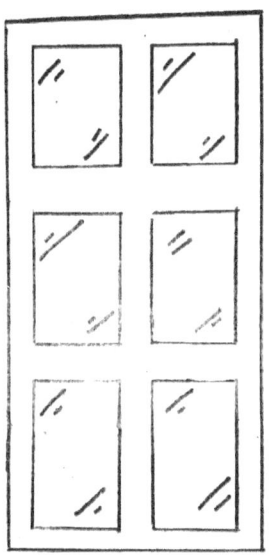

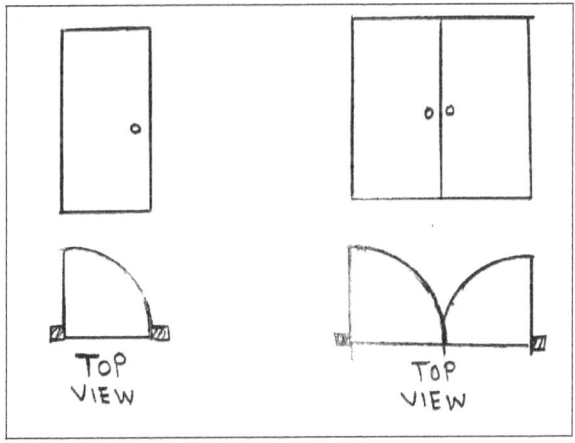

- **Inner Doors**

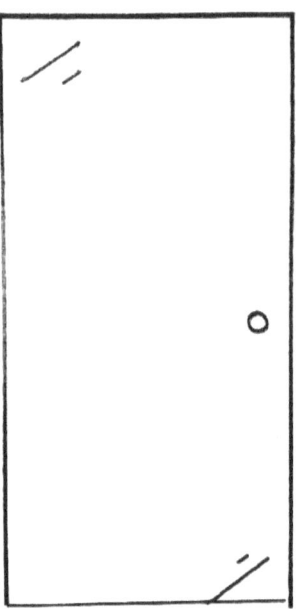

The doors used for inner parts of the house are mainly flush door single shutter. Flush door is a completely smooth door made of plywood of MDF (Medium Density Fibre) over a light timber frame. The skin of such doors is made up of hard boards.

- **Bathrooms and WC Doors**

The doors used for bathrooms and WC are UPVC doors (Unplasticized Poly Vinyl Chloride).

❖ Windows

Windows allow natural light into the rooms during the day and let air to keep it fresh and dry. Also provides a view to the outside of the room.

The sill level (lower level of window) should be kept at a height of 0.75m to 1m from the floor level. While the lintel level (Upper level of window) should be at 2.1m height from floor level.

- **Sliding window**

It has horizontal sliding sashes operated by sliding along a track in the window frame.

- **Hinged window**

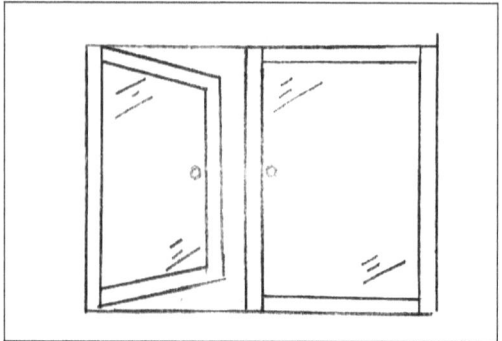

The sashes are hinged from one side and opens like door in one direction, generally in outside direction.

Chapter 17 – Completion/Occupancy Certificate.

Obtaining completion certificated is the last step of the process.

Certificate or Occupancy certificate is the documentation declaring that the house is in ready position and is fit for habitation. The certificate is the assurance that the house is constructed by the building laws and regulations, and is in suitable condition for occupancy.

The developer has the responsibility to obtain the OC (occupancy Certificate) from the capable authority of the area.

Failing to obtain this certificate may lead to serious problem in getting access to civic amenities like electricity, water supply etc.

---THE END---

www.ingramcontent.com/pod-product-compliance
Lightning Source LLC
Chambersburg PA
CBHW031626210526
45464CB00004B/1765